I0040879

BEEKEEPER'S COMPANION

ILLUSTRATED FIELD GUIDE

BEEKEEPER'S COMPANION

ILLUSTRATED FIELD GUIDE

Adesina Daniel Oduntan

IC
PRESS
Idea Creations Press
www.ideacreationspress.com

IC
PRESS

Idea Creations Press
www.ideacreationspress.com

Copyright © Adesina Daniel Oduntan, 2018.

All rights reserved. No part of this book may be reproduced or transmitted in any form or by any means, electronic or mechanical, including photocopying, recording, or by an information storage and retrieval system, without permission in writing from the author.

978-1-948804-04-2

Publisher's Catalog-In-Publishing Data

Oduntan, Adesina Daniel, author
Beekeeper's Companion – Illustrated Field Guide / Adesina Daniel Oduntan
First trade paperback original edition. | Salt Lake City: Idea Creations Press, 2018.
ISBN 978-1-948804-04-2 | LCCN 2018950753
Bees | Beekeeping | BISAC: Beekeeping Field Guide

Printed in the U. S. A

Table of Contents

Acknowledgements

I can never forget how my journey into the beekeeping world commenced. The vessel (man) that God positioned for my start-up, a brother, senior from my Alma-mata of Baptist Boys' High School, Abeokuta, Nigeria, was someone who believed in me and gave me all needed financial support in 1993, in the person of Elder Olusola Abraham Adeyemi, retired permanent secretary at Ogun State government civil service. He is my mentor, my role model; the man that I always admire to imitate and possibly to share his traits of giving ability, and a man with listening ears. I owed him so much appreciation, that without him my "DREAM" would not have become a "REALITY".

Here comes a man that opened my destiny and connected me to my destiny helper, the man is my teacher, mentor in person of Dr. Stefan Stangaciu. I came across Stefan in my search for knowledge to know more on anything regarding "BEES". I did my Apitherapy Internet Course (A.I.C) in 2006 through Stefan Stangaciu and he invited me to German Apitherapy Congress in Germany in 2007, and joyfully handed me over to Arno Bruder, my God-sent friend, brother, mentor, teacher, who taught me all I know in beekeeping today. Arno created so many opportunities for me to be all I could be.

I am very grateful to this "TRIO". God established them for me as the bedrock of all I am today.

My sincere appreciation to my wife Grace Motunrayo, who has been my bedrock and supports and my children Daniel Opeyemi, Elizabeth Oyinlola, and Glory Feyisayo for their supports and understanding through thick and thin of our journey to where God has brought us.

My heart-felt gratitude to Prof. Floyd Ostrowski, Sarah Idehen (President of Nigerian Association of Utah), Kathryn Jones, Douglas Jones, and Prof. Kolawole Adebayo for their support to proof read this manuscript.

I am appreciative to members of Utah Microenterprise Loan Fund (UMLF) and International Rescue Committee (IRC) Utah for their financial support to start my business here in United States. My special thanks go to these wonderful ladies. They really play an important role in my business here in Utah, and they are Shailaja Akkapeddi, and Jane Ullah, also my good friend Rogelio Nava, who designed my website. I cannot forget Matthew Wallace who upgraded my website, and Natalie Shahmiri who designed my book cover and my company logo, and my fellow beekeeper and friend who supported me for my first 5 colonies in Utah State, Lee Knights of Knights Family Honey.

Finally, my recognition goes to the host of many other people who have in one way or another assisted and contributed in no small measure to making this work a reality.

Adesina Daniel Oduntan
July 2018. Utah, USA.

Preface

The economic situation in the world in general has warranted serious attention, especially in the developing countries, and this has called for entrepreneurial skills development and innovation in every professions, as the saying goes that whatever your profession …farm (Bee farming especially).

Agriculture plays a prominent role as a source of raw materials for many value-added products chains. Beekeeping has played an important role in reviving the nation's agricultural potentials and rewarding the activity and knowledge of enterprising individuals and generating self-reliance.

Bees are nature's gifts from God to help mankind for our survival. It is a known fact that one third (1/3) of world food population is aided by bees through pollination. Despite the pollination services offered by bees, there are nine (9) other beehive products derivable from keeping bees, which are Floral honey, Honey-dew honey, Bee pollen, Bee propolis, Bee royal-jelly, Bee bread, Bee apilarnil, Bee venom and Beeswax.

Many people are scared and afraid of bee stings and regard bees as little deadly creatures, despite the benefits bees provide in nutrition, health and wellness.

This book is primarily designed for beekeepers, both the hobbyist and professional, to know what to do to make the best advantages of bees to improve their livelihoods. This book is an illustrated field guide that shows you step-by-step activities to be done at the apiary. It not only shows you what to do, but it gives you an illustration with pictures to do it right the first time.

Chapter one of the book explains the biology and ecology of bees, how to start right with bees, floral chart calendar and seasonal management of the bee colony.

Chapter two and three (2&3) delve into necessary tools in beekeeping for efficient and effective productivity, how to help bees experience their natural environment through proper manipulation of the colony and understanding of the basic principles on what to look for in the colony for good management.

The chapter four (4) is an interesting part of the book that may fascinate the beekeepers. It shows how to harvest all the beehive products to fetch additional income, improve their wellness and nutritional value of beehive products. These are well illustrated with the pictures.

Chapter five (5) explains the queen rearing in a simple way that a beginner beekeeper can raise his/her own queen bee with or without the support of a mentor.

The last chapter, which is chapter six (6) explains in detail what it entails to start your own business, the factors to be considered.

Finally, this book is written to encourage people out there to not wait for things to happen to them, but they should make things happen for themselves.

Adesina Daniel Oduntan,
e-mail: vocationalbeecraft@gmail.com
www.beecraftconsult.com
July 2018. Utah, USA

Foreword

The beehive is a super organism where all the individual members dedicate their lives for the greater good of the colony. I believe Adesina Daniel Oduntan is imitating bee behavior in his selfless endeavor to improve the practice of beekeeping, and to use beekeeping to improve the quality of life of the impoverished by teaching them the beekeeping business skills. He is one individual whose primary focus is the betterment of the "hive".

I met Daniel several years ago when he was teaching beekeeping in Nigeria. His passion for the well-being of God's greatest creature – the Honey Bee, combined with his passion for building an industry where the underprivileged could participate, is inspiring. He has helped many families improve their lot in life by teaching the correct way to manage bees.

Beekeeping changes people's lives. Becoming a caretaker to this lifeform is a huge responsibility and crucial for the survival of mankind. Yet the benefits of such work will provide much required income for the sustainability of family life in even the poorest regions. Daniel recognizes these commitments and this book will help guide the neighborhood hobbyist, as well as provide the education necessary for a family to make a living wage.

This guide should accompany every beekeeper in the field. There are quick references and information on dealing with virtually every situation. The easy to read and quick access makes this guidebook a must for every beekeeper – beginning and expert.

Many folks consider St. Valentine to be the Patron Saint of love and romance. But he really is the Patron Saint of Beekeeping. St. Valentine is said to "ensure the sweetness of honey, and the protection of the beekeeper". Daniel is the modern-day St. Valentine.

Professor Floyd Ostrowiski
Former Vice President of Operations,
A.I ROOT Company,
Medina, Ohio State.
United States of America

Dedication

This work is dedicated to the glory of the

ALMIGHTY GOD

And the

BENEFIT OF MANKIND

And

To the Honor of

My dear MOM

FELICIA IYADODE AJIUN ODUNTAN (nee KILANKO).

Chapter 1 - The Biology and Ecology of Bees

Colony & Organization

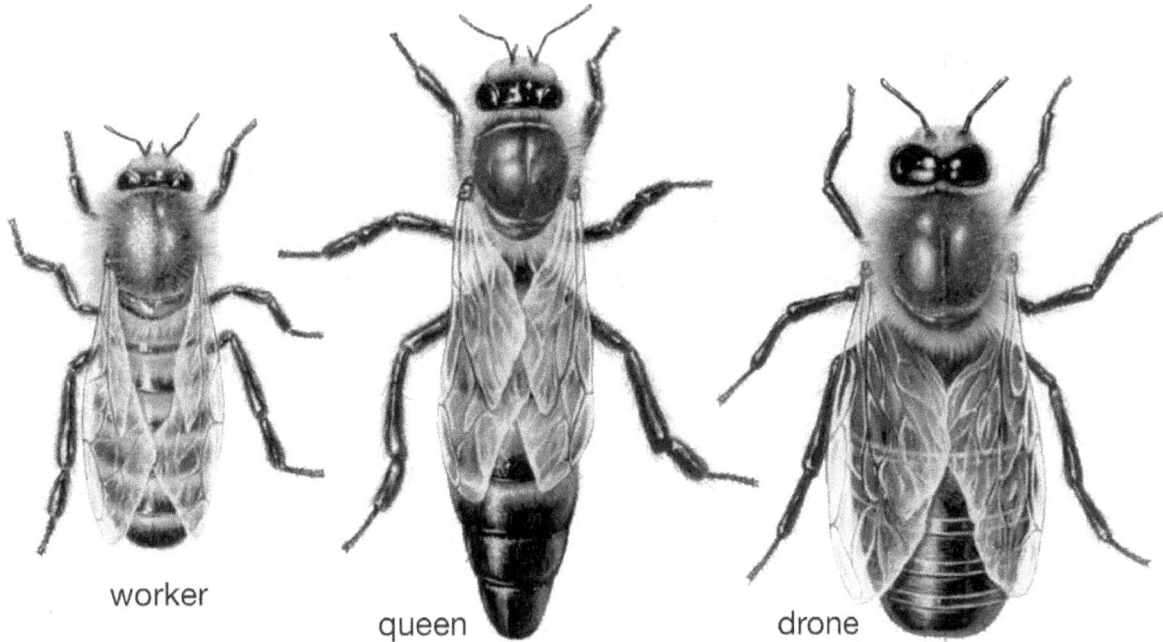

worker   queen   drone

Honey bees are social insects that live together in a large, well-organized society as a **COLONY**. They engage in a variety of complex tasks including communication, environmental control, defense, and division of labor. There are 3 castes in a colony: **WORKERS**, **DRONES**, & **QUEEN.** Individual bees (workers, drones, and queens) cannot survive without each other (colony).

Source: (picture) yourlandprotective.com

Colony of Bees

As a beekeeper, one must accustom and familiarize oneself with the understanding of the nature and behavior of bees by gathering some insight into the society of human beings. Both human beings and bees live in a highly complex society, i.e. the being in each society has specific jobs and they must contribute to the well-being of the entire population.

Queen Bee

A colony has only one queen, the only sexually developed female. Her primary function is reproduction – the laying of eggs (300-2,000 in a day). The queen leaves the hive only during swarming and mating periods.

The characteristics of the colony depend largely on her genetic makeup – along with that of the drones she has mated with – which contribute significantly to the quality, size, temperament, and productivity of the colony.

Worker Bee

Workers are sexually undeveloped females. They have some specialized structures, such as wax gland, pollen baskets, brood food glands and scent glands. The works they perform are related to their age. There are in-house bees and field bees. The first are responsible for the building of the comb, the cleaning of the comb, and feeding the young brood and honey ripening. Field bees collect water, pollen and nectar, as well as plant sap resin (propolis).

Drone Bee

Drones are male bees, only present during late spring and summer. They have no stinger and hence do not harm. They are only to fertilize the virgin queen during her mating flight

Honeycomb

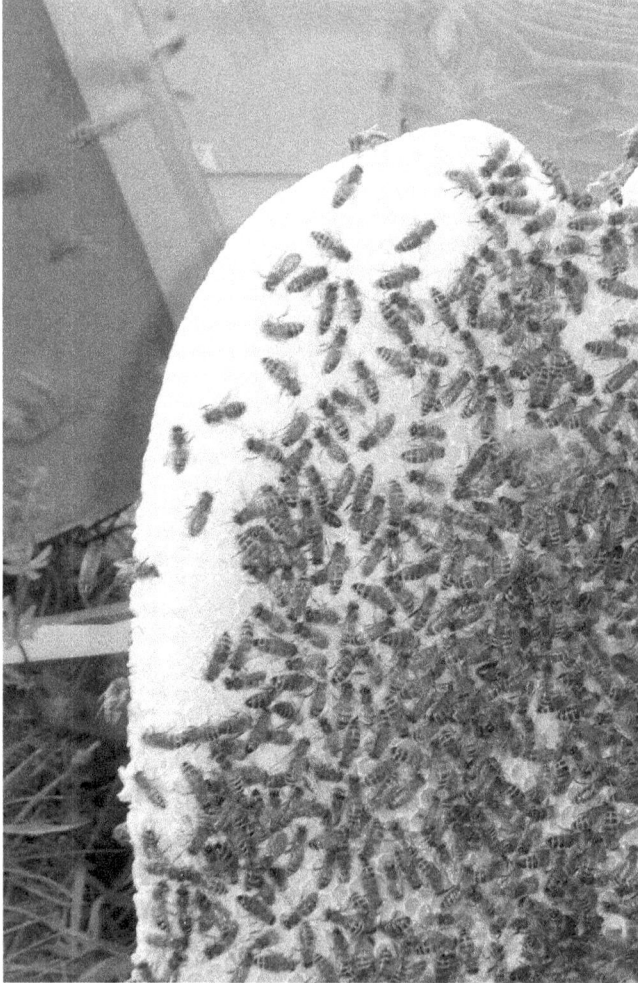

The mass of hexagonal cells of wax was built by honey-bees in which they rear the young (brood), store honey, and deposit pollen.

The mass of hexagonal cells of wax was built by honey-bees in which they rear the young (brood), store honey and deposit pollen.

It is the age of the colony that determine the color of the comb and the honey. The darker the color of the comb and the honey, the older the colony age.

Cluster of Bees

A group of bees looking for a new house.

Floral Honey

Floral honey is the natural sweet substance produced by the honey-bees from the nectar of plants or from secretions of living parts of plants, which honey-bees collect, transform by combining with specific substance of their own, deposit, dehydrate, store, and leave in the honey-comb to ripen and mature.

Honey that comes from the nectar of plant is called BLOSSOMED HONEY or NECTAR HONEY.

Source: Beekeeping in the United States

Honey-Dew Honey

Honey comes mainly from secretions of plant sucking insects (Hemiptera) on the living parts of plants or secretions of living parts of plants.

Bee Development

A good beekeeper must inspect the beehive curiously to understand the bee development, but a beehive should never be approached without some goals in mind, especially to see the QUEEN. She is hard to find at times.

The term "brood" is commonly used to designate the young bees that have not emerged from the cells. The larvae (or even the eggs) are in various stages of growth. Sometimes, the beginner is confused because he is not able to distinguish capped honey from capped brood, nor does he know the difference between drone and worker brood.

Unfertilized eggs become drones, while the fertilized eggs become either workers or queens. Nutrition plays an important part in caste development of female bees. Larvae destined to become workers receive less royal jelly and a greater mixture of honey and pollen compared to the copious amount of royal jelly that a queen larva receives.

Eggs   Larva   Pupa   Adult

The developmental stages are similar, but they do differ in duration.

Eggs

It is the presence of eggs or young larvae that shows that the bees have a queen and are beginning to rear brood. On the other hand, the absence of unsealed brood and especially the absence of eggs may be an indication that the colony is queen-less. During spring and early summer, there will be or should be brood in all stages, including eggs. Such a condition indicates prosperity and a beekeeper can feel that his bees are doing well. But if there are no eggs nor young larvae, and the queen cannot be found, if there are also initial queen cells, the possibilities are that the queen has recently died or that a swarm has ensued.

Drone Cells

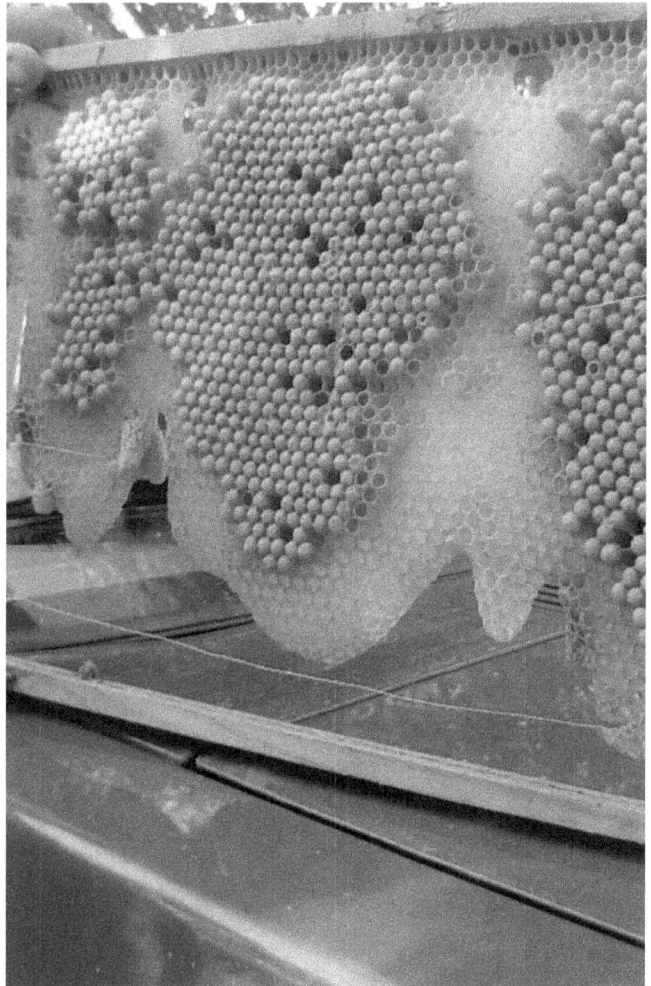

Sealed brood of the worker or drone is of light to dark brown color, depending on the age and color of the comb itself. In ordinary worker brood, the capping is made of wax materials, smooth and slightly convex if the brood is not diseased. Drone brood is the same in appearance except that the capping is more convex with four cells to the inch.

Worker Cells

The larger capped cells with the rounding heads are above the surface of the comb. At the top, bottom and cover of the combs are drone brood. Worker brood, on the other hand, which covers most of the combs, is flat on top and smaller in size. Worker brood, roughly speaking, has five cells to the inch, while drone is only four cells to the inch.

Queen Cell

If there are no eggs nor young larvae, and the queen cannot be found, and if there are also initial queen cells, the probabilities are that the queen has recently died or that a swarm has ensued. It may further be said that the absence of egg and the presence of the initial queen cells during the active season are almost absolute proof that the queen is not in the hive or that she is about to be superseded.

Swarm Cell

Swarm cells are initiated in the colony for the purpose of leaving to get a more spacious location as a result of lack of space. This makes it impossible for the worker bees to build more combs for queen to lay eggs.

The virgin queen will emerge from the cell to replace the mother queen after the "Prime Swarm" has left the colony together with adult bees and an abundant honey supply.

It is an opportunity to establish another colony provided you are fortunate to discover the swarm cells before the bees leave the colony. When you do not want to divide the colony, you destroy the swarm cells and create more space for the colony.

Getting Started Right with Bees

There are several different ways of getting started in beekeeping: buying package bees; purchasing a nucleus (nuc) colony; buying established colonies; collecting swarms and taking bees out of wall cavities.

Package Bees

Package bees are available in a 3-pound size. A newly mated queen is included to be used for developing new colonies. Package bees should be ordered in January to ensure timely delivery in early spring (April). Package bees are sensitive, handle them with care. Before installing them, protect them from wind and cold, but do not put them in a heated area.

You should install package bees as soon as possible after their arrival, you cannot delay the installation more than 48 hours without encountering problems. Feed the bees immediately and continue feeding them until they have stabilized.

Nucleus (nuc) Colonies

A nucleus (nuc) is a smaller hive consisting of bees in all stages of development, food, and a laying queen, including enough worker bees to cover from 3 – 5 combs. Supplemental feeding should be given immediately when it is transferred to the full-size hive body. The advantage of a nucleus colony over the package bee is that faster colony development is achieved because of the presence of brood, and no break in the queen laying occurs; still, this option is more expensive than the package bees.

Buying Established Colonies

Buying an established colony is not advisable for a new beginner, but experienced beekeepers may find this a practical means to increase their numbers of colonies. Purchasing smaller units such as package, or nucs in the spring allows a beginner to grow in confidence and managerial skills as the colony size increases during the season.

Collecting Swarm

Swarms normally cluster on a tree limb, shrub, fence post, or on the side of the building. When swarms are located, spray sugar water or ordinary water on them to keep them calm; do not use smoke. Remove the swarm gently. Put it directly into the hive or enclosed box. If the swarm cannot be cut down, either shake or scrape the bees into the lightweight box.

In the tropics region, beekeepers relied heavily on catching the swarm of bees by hanging the 5-frames swarm catcher box on the trees and transferring them when it is colonized.

Once you capture a swarm. Introduce them into your own hive by either shaking or dumping the bees into an open hive with drawn combs, which is better than foundation; one or two drawn combs preferably with pollen. Brood and or/ honey combined with foundation works great.

Bees Evacuation from the Rooftop of the House

The best way to remove a colony from a wall or roof top is to remove the exterior coverings to completely expose the colony. Then cut out the combs and brush or vacuum the bees from the interior of the wall. You can then hang your baited hive nearby for a swarm catch or collect the swarm from nearby when they are all settled down from looking for a new home.

Do You Know What to Do?

As a beekeeper, try to visit your apiary 2-4 times a month. Take a closer look and properly observe the foraging traffic activities of the bees in a sunny day. Carefully study the in-house bees as well.

Before the purchase of beekeeping equipment, you must know where to locate your apiary. There are many factors to be considered before you start off, and it is advisable to start small.

Beekeeper's Role

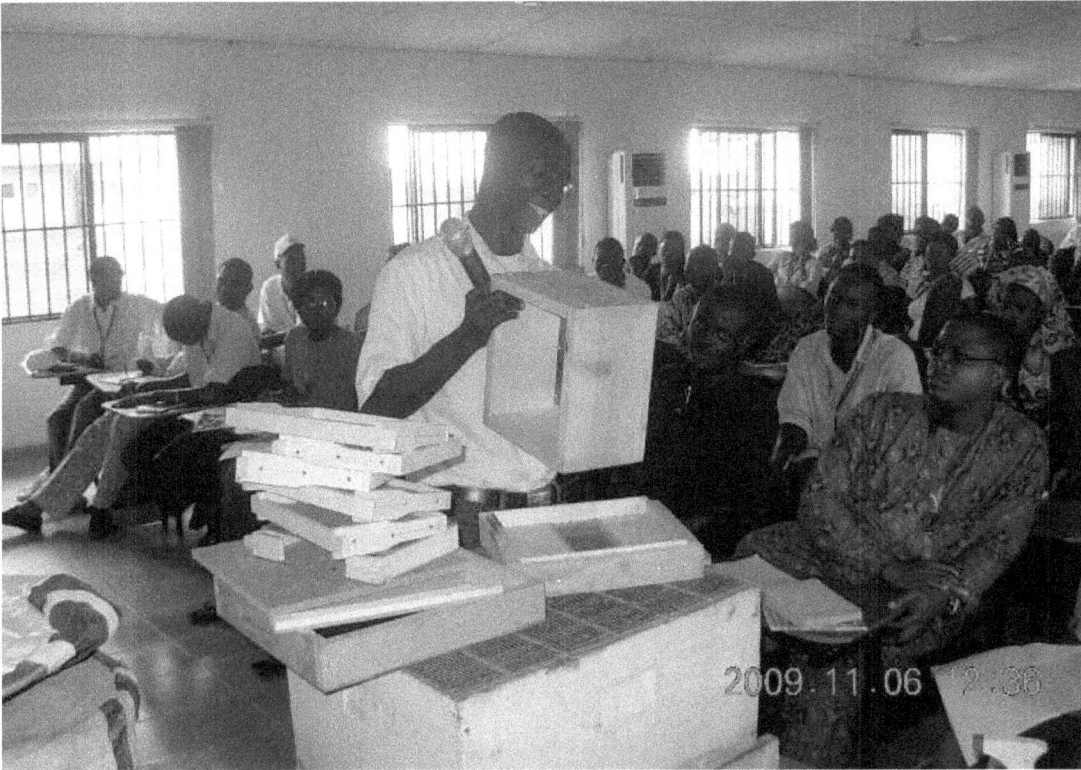

For a successful beekeeping career:

Seek to further your study on bees by attending beekeeping training classes.

Make a friend with a beekeeper in your neighborhood.

Buy books on the beekeeping subject.

Register with local beekeeping association near you.

Attends seminars and beekeeping workshops, and finally,

Invest to buy beekeeping tools for your own apiary.

Apiary Location Requirements

Before the siting of the colonies, consideration should be given to the following factors to assess the suitability of the site for keeping bees.

Criteria for the site are as follows:

Is there adequate forage in the surrounding area for the colony to support itself and obtain nectar and pollen throughout the year?

There must be no question of danger to human beings, particularly children, or animals.

Ideally, the apiary should be in a place where nobody except the beekeeper can be stung.

Source: Beekeeping Study Notes -- J.D. & B.D. Yates.

Under no circumstance should an apiary be established adjacent to a public thoroughfare, even if there is a barrier (e.g. hedge or wall) of suitable height between.

The site should not be in frost and protection from the prevailing winds is most desirable.

The site should be accessible by road throughout of the year.

The site should be surrounded by a stock-proof fence, if it is adjacent to pasture where livestock is likely to be grazing.

Source: Beekeeping Study Notes ---J.D. & B.D. Yates.

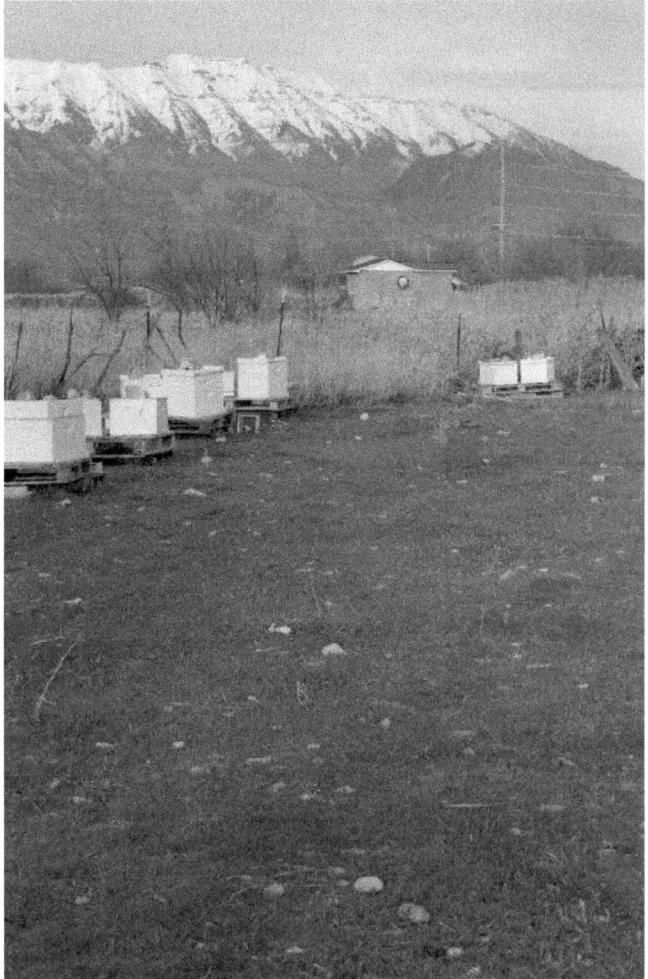

The factors to be Considered in the Siting of Colonies in Home and in out-Apiaries.

General considerations when siting both home and out apiaries are as follows:

Stocks must be sited so that the flight path of the bees avoids foot paths and areas where there is likely to be any human or animal activity. Stocks can be sited so that bees must fly up and over hedges and fences thereby getting the bees to a safe height above anyone on the ground.

There must be plenty of space around each stock for colony manipulations and maintaining the site (e.g. grass cutting).

The layout of the stocks should be in an irregular fashion to minimize drifting.

Bees in the stocks will at some time swarm despite the best efforts of the beekeeper to prevent this happening. Shrubs and trees around the stocks are useful for the swarms to hang on.

Source: Beekeeping Study Notes ---J.D. & B.D. Yates.

Plants and Bees

A list of major nectar and/or Pollen producing plants and their flowering periods in your locality should be a companion.

Every beekeeper is expected to be familiar with the vegetation near/or round his/her apiary for a successful beekeeping operation. Understanding the major nectar and pollen producing plants and their flowering period is of great advantage to maximize the efficiency of the bees. It is of great importance for a successful beekeeping operation to keep floral calendar charts and plants for bees where necessary.

Source: Beekeeping in the United States.

2008.12.15

To identify nectar – producing plants i.e. honey source, detail information must be known and quantified, where necessary, of the plant, its economic uses, its flowering period and nectar (or honey dew) flow as well as its pollen and its honey production period and of the honey's chemical composition and physical properties including flavor, and granulation.

It is important to keep a record on a floral calendar for proper monitoring of the nectar flow season within one's own apiary; useful data on pollen plants must be studied and a table chart must be prepared for proper understanding of each plant.

Source: Beekeeping in the United States.

The following criteria for the selection of plant and honey characteristics will be found useful; botanical names of plants, family and common names must be noted. Vegetable forms of plants i.e. floral descriptive, distribution and habitat should be of paramount interest, economic and other uses as well as nectar rating and its blooming period are also of great importance.

Source: Beekeeping in the United States.

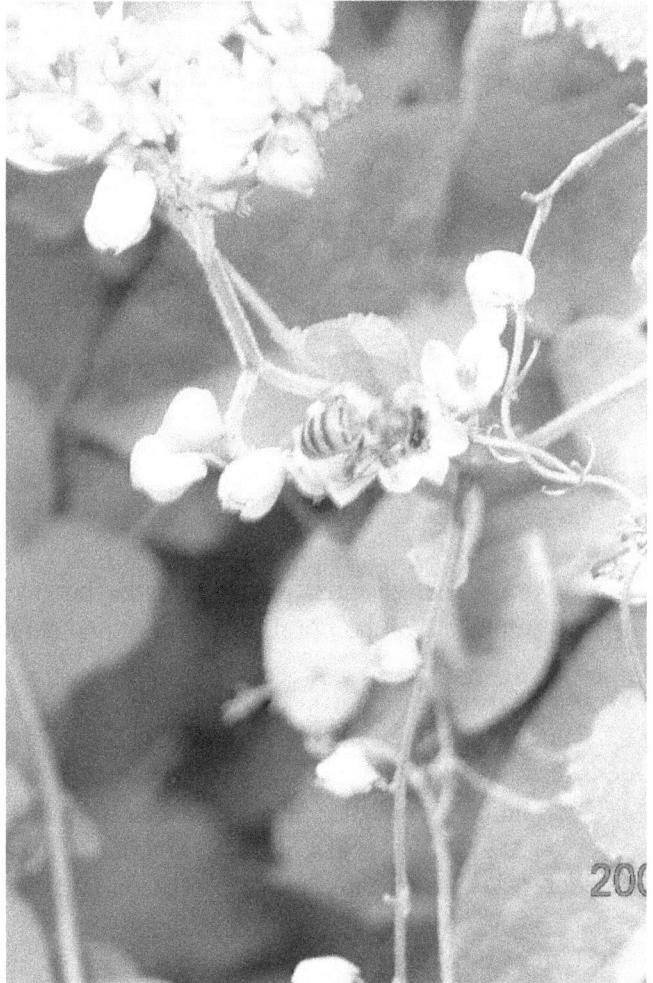

Flowering Chart Calendar (FCC)

Names of Bees Tree	Jan	Feb	Mar	Apr	May	Jun	Jul	Aug	Sep	Oct	Nov	Dec
Mango												
Citrus												
Cashew												
Sunflower												

A Floral chart calendar determines the botanical and geographical origin of honey, especially the characteristics of the floral distribution and density within an apiary.

The chart reveals the buildup and nectar flow periods and the availability of nectar and pollen in the vicinity for successful honey production.

Such information also enables a bee farmer to determine when to install packaged bees, divide colonies, put on supers, use swarm control measures, remove honey, re-queen, prepare colonies for nectar flow season and locate profitable apiary sites.

Source: Beekeeping in the United States.

The Seasonal Management Calendar for Beekeepers

	TEMPERATE	REGION	
Dearth season WINTER	Build-up season SPRING	Nectar-flow SUMMER	Honey-flow FALL or AUTUM
NOV/DEC---- FEB/MAR	APRIL-JUNE	JUNE-AUG	SEPT—NOV
Honey flow or Harvesting season	Dearth season	Build-up season	Nectar-flow season
	TROPICAL	REGION	

Keeping bees profitably requires the ability and knowledge of a good beekeeper to understand the weather in which the bees are striving together with the method of bee management techniques employed. A beekeeper's year is divided into four (4) phases as follows:

1. Dearth season, 2. Build-up season, 3. Nectar-flow season, 4. Honey-flow season.

The Dearth season (April-June)

This period affords us the opportunity to capture bees from feral colonies into our empty hives. It is the best time for a new farmer to start bee farming. Regular inspection against pest infestation especially wax-moth and feeding of sugar syrup should be done.

Among serious problems at this time is that bee population is drastically reduced due to nectar-scarcity. Absconding resulting from weather fluctuation, bees are usually aggressive and spend time making propolis. Strengthening of weak colonies and a hive's temperature maintenance are activities carried out at this time, as well as the building and repairing of hives, and the reduction of hive entrance size to aid security from intruders. Records keeping, planning and budgeting for the honey flow is necessary.

This is the busiest time in the temperate region to manage the bees.

Build-up season (June-August)

Most of the new colonies are established and have commenced brood rearing. Both the new and old colonies now grow in strength. This is the time to take care of colonies for early swarming. Amalgamation (uniting) of the weak colonies and initiation of queen rearing and looking out for swarm preparation and prevention is the major pre-occupation.

This is the dearth season in the tropical region when bees are to be fed with sugar syrup because it is the rainy season when the pollen and the nectar in the plant would have been diluted or dropped off the plant.

Nectar-flow season (Sept-Nov)

Bees are active and the honey in the hives are unripe at this time. The size of the hive entrance should be increased to avoid delay of flight activity during foraging. Regular inspection is the routine during this period. The period to avoid swarming, because it is the major problem at this period mainly due to lack of space in the hive to lay eggs and deposit nectar.

In the temperate regions, major activities are health treatment and provision of sugar syrup for the winter time.

Honey-flow/Harvesting season (Dec-Mar)

Many Top-bar hives are filled with ripe honey. Pollen and nectar intake reduces, and entrance activity decreases; but inside activity continues. Sample harvesting is done in December. While major harvesting is carried out in January to March. More promising colonies can be harvested twice after the first harvest. Honey and wax will be ready for market after the extraction and processes. In the temperate regions no activities with bees whatsoever because of weather condition (winter season).

Things to do to Establish a Good Apiary

FOR DEVELOPING
COUNTRIES (Kenya Top-bar hive).

Make a recognizance survey of less than half a kilometer around the vicinity of your apiary to study the bee plants. Take picture of bee plants, weeds, shrubs as well as tree crops visited by bees.

Write your list of bee floral plants and draw a floral chart calendar to study the flowering season (nectar secretion) of each plant.

In developing countries, we depend so much on swarms of bees to stock our hives. Hang the swarm catchers/boxes (5 Frames hive) for a swarm of bees at different locations, and when the box is colonized (caught bees) – transport them to your permanent apiary and transfer them to your normal hive to continue development.

A good suggestion for successful establishment of an apiary mostly in developing countries, is for a beekeeper to be very familiar and friendly with their host community (farm settlers). He should get the villagers involve and encourage them to adopt beekeeping practices through training and provide them with job opportunities as a security measures to avoid thefts and vandalization.

FOR DEVELOPED COUNTRIES

Package bees are sold to stock beehive in advance countries like the USA, Europe, Australia, & China. Package bees are produced by a set of beekeepers called QUEEN BEE BREEDERS. They do not need to rely on a swarm of bees to start beekeeping, but they can take advantage of a bee's swarm to increase their colonies in the spring.

There can be theft issues in some places in advanced countries as well, but these issues is not as rampant as those experiencing in developing countries. In some places where farm owners lend you a portion of land to keep your colonies, you can show appreciation in return with some bottles of honey.

Hiving Swarm/ Swarm Catcher

If a swarm is in inaccessible place, use a swarm bag with a long pole to reach out to them. Collect them and transfer them immediately to the swarm box below.

Hiving a swarm of bees is a process whereby the naturally built cluster is collected or caught into the hive directly.

A swarm can be collected through a swarm catcher hanging on the tree branches in the bush. If the bees cluster is on a limb of a tree, the simplest way of hiving or collecting them is to first spray sugar water or ordinary water on them to keep their bodies wet and to prevent them from flying. When their bodies wet, they will hold on to each other, and you can then cut off the limb above the clustered bees and carry them to the hive.

To secure the swarm of bees permanently in the hive, put 1 or 2 brood frames of bees, with stores (honey frame) and feed the swarm immediately with sugar syrup.

In a case where feral colony is in an inaccessible location. Climb a ladder to the location and destroy the colony. Cut off the fixed-comb produced by the bees to collect the swarm.

Feeding of the package Bees/ Bees Swarm

Immediately after the package bees are installed or bees swarm is collected, feeding of the bees with sugar solution is very necessary and important for sustaining the colony before they continue from there for foraging.

Beehive Positioning & Arrangement

Beehives can be randomly arranged at the spaces of 10- 20 meter apart in an open field or in a forested farmland.

Beehives can be randomly arranged at the spaces of 10- 20 meter apart in an open field or in a forested farmland.

Urban Beekeeping

Keeping bees in the cities must take several precautions to decrease the chances of your colonies becoming a public nuisance. Maintaining gentle colonies is imperative in highly populated areas. Most of the bee plants in the city walk way is of advantage to the bees kept in the city.

Keeping bees in the cities must take several precautions to decrease the chances of your colonies becoming a public nuisance. Maintaining gentle colonies is imperative in highly populated areas. Most of the bee plants in the city walk way is of advantage to the bees kept in the city

Backyard beekeeping is a good starting point for a hobbyist beekeeper. It gives support to the local gardens and sustains the community of bees.

Apiary Sanitation

Good apiary hygiene is a must.

Do not throw brace comb on the ground, always place it in a suitable container.

Always keep the apiary clean and tidy.

Arrange all hives in such a way that drifting is reduce to a minimum.
This mechanical grass cutter is use for clearing and cutting of the grass in the apiary.

Chapter 2 – Hands-on Demonstration Class

Adesina Daniel Oduntan

Honey-Hunters

In developing countries, bee-keeping is fast
becoming a traditional occupation. In some
areas, men who have no hives, raid the colonies
of bees in natural habitats when they are
discovered.

Source: (picture) Smithsonianmag.com

Tools for the Bee-Keeping Job - Straw – Hive

In an effort to get honey by the traditional beekeepers, various types of hives are used; examples are straw hive, pot hive, and gourd hive. This kind of beehive is not accessible for inspection.

Bees produce "fixed-comb" which can not be inspected until harvest time is due, and the comb must be destroyed to harvest the honey. Depending on the location, the amount of honey and types of honey varies.

The local craft-men produce this kind of beehives with treated grass and earn a little money from the sale to local/traditional beekeepers.

Tools for the Bee-Keeping Job – Clay Pot Hive

Traditional bee-keepers are often highly skilled. They know exactly where to place hives and when to harvest them. The honey crop is obtained by plundering the bees nest, sometimes resulting in the complete destruction of the colony.

Beekeeping equipment needs/requirements vary with the size of your operation, types of beehive to use, numbers of colonies to manage, and the type of beehive products you plan to produce.

The basic beekeeping equipment: comprises of the beehive (Top-bar hive, or Langstroth hive), bee-suit, smoker and hive tool, honey fork and honey extractor for handling the honey crop.

Tools for Beehive Assemblage

The tools for beehive making includes:
  Wood body of difference sizes
  Wood frames
  Air compressor
  Wood glue
  Clapper
  Hammer
  Nail gun machine
  Pins
  Jig for frames
  Square ruler

Beehive Assemblage

The beehive assemblage requires little or no technical know-how of carpentry skills. The tools and equipment are available at reasonable cost, and any interested beekeeper can afford to buy what he/she needs to assist himself/herself to realize their beekeeping dream.

Beehive – Langstroth Hive

The wooden beehive is a man-made structure in which the honey bee colony lives.

Langstroth hive: A typical Langstroth hive consists of a hive stand, a bottom board with entrance reducer. Hive bodies with suspended frames containing foundation or comb, and inner and outer covers.

The hive bodies that contain the brood nest may be separated from the honey supers with a queen excluder. This is the most widely used commercially in the world. The frames are separated from the hive wall (and from each other) by a bee space.

Most Langstroth hives have boxes to accommodate 10 frames, but 8 and 12 frames are also made. The modern beehive is like a highly efficient multistoried factory with each "story" having a special function. These "stories" work together to provide a home for bees, and a honey factory for beekeepers.

Kenya Top Bar Hive - for Tropical African Bees

The hive internal measurements are 74.4 cm by 47 cm at the top and 74.4 cm by 24 cm at the bottom and a height of 26 cm. It has 22 top-bar frames with 3.2cm wide, 48cm long, the top-bars touch each other and there is no space between them.

 This is an important feature when handling tropical African bees. Each bar must be taken out one at a time. Smoking through the opening reduce the flight and stings of the bees.

Hive Tool

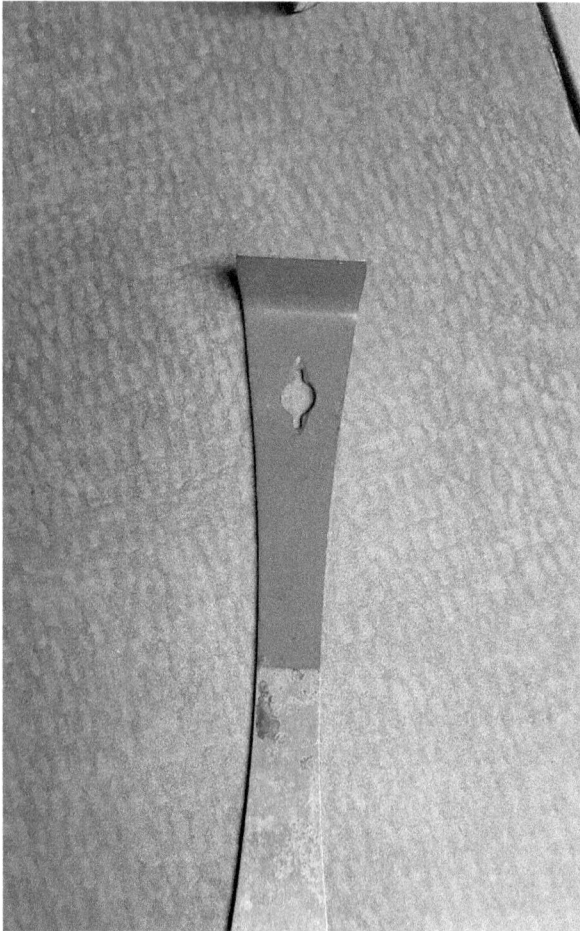

This is a strong metal bar about 250mm (10inches) long. One end is chisel-shaped and can be used for separating boxes glued together with propolis and for loosening frames. The other end is slightly turned up and sharpened for scraping propolis and wax off wooden surfaces.

Protective Clothing - Bee-Suit)

A beekeeper should be adequately protected from a bee sting as this may mar his working on his colonies. Different colonies exhibit different characteristics and working on too many colonies at a time without adequate protection can be suicidal. Modern fastening devices, such as zip fasteners, have made it possible for a bee-keeper to be completely enveloped in a single garment keeping them safe.

Smoker

The first piece of equipment necessary when thinking about opening a hive is the smoker. The smoker consists of a metal fire pot and grate with bellows attached. These give a directional flow of smoke. Smoke has been used as a pacifier all over the world for thousands of years. In traditional beekeeping, smoldering twigs or grass are used but this does not give the directional flow of cool smoke that is most effective and best for keeping the bees quiet. Fueling (smoking) materials include cow dung, papaya fibers, and dried grass.

Source: Beekeeping in the United States.

A smoker is a beekeeper's companion. When a smoker is not in use, the bees are in charge, but if you smoke them, you are in control. Smoking the bees while operating on them secures you added protection. The bees will fly away from smoke and busily engorge themselves on honey instead of stinging you.

How to Light a Smoker

Take a sheet of paper, crumple, and light with a match and drop into the smoker. Pump the bellow several times to keep it burning. Place a bunch of smoker fuel (cow dung, dry grasses) to be used in the smoker tank loosely and pump the bellow briskly to ignite the fuel and get it burning. Open the tank to add more fuel materials when the smoke is going out.

Top Feeder

In most parts of the world, it is necessary to make provision for the bees by supplying sugar syrup in a feeder.

To counteract some unexpected adverse weather (dearth season).

To build up a small nucleus made to increase the number of colonies.

To encourage a swarm, or bees newly put into a hive (package bees) in the form of a dummy frame or top hive feeder.

Dummy Frame Feeder

This is sometimes called a division board feeder; it conforms to the size of the frame-plus-comb in the brood box, where it replaces a complete frame. The bees enter from the top, and inside there is a float or some other provision to protect the bees from drowning in the bulk of the syrup. This feeder is safe from robbing, and its content is easily available to bees, but the hive must be opened to fill it.

African beekeepers can make this feeder using top-bar hive frame. Cover both sides of the frame round. Then, cut the bar close to the handle at both end. Make sure the upper part facing up is opened for the bees to reach for the sugar syrup inside the feeder.

Feeder

Large feeders are preferable for food that the bees must store for a dearth season when it is important that the bees take the syrup immediately. Dry sugar can be fed instead of syrup in cool weather; spread over the inner cover of the hive, it needs no special feeder. Dry sugar feeding will not lead to robbing, unlike syrup or honey feeding if other bees have access to the food (most especially with an outside feeder).

Solar Beeswax Extractor

Beeswax is a valuable hive product, and it brings the beekeeper an added income. Unlike honey, it needs no container, and no special care even in long term storage. Despite this, it is all too often thrown away. When beeswax capping, combs, etc. have been washed free from honey or have been cleaned by bees, the beeswax is melted to let everything that is not pure beeswax separate out by sinking to the bottom.

Steam Beeswax Extractor

It is essential that light combs should be treated separately from dark ones, because light wax will fetch the highest price. Beeswax must be heated in a safe way or there is danger of fire.

The beeswax capping should be put in this capping spinner to melt the wax and separate the honey in good condition and settle it down below. The wax will float to the top.

Propolis Trap

The principles of harvesting are simple. A flat sheet (wooden or metal) with width slits of 6.0mm that bees will close with propolis.

The trap is inserted at the top of the hive, where bees regarded it as opening and work to close it with propolis. Remove the sheet and place in the freeze. The propolis can be released from the sheet by shattering.

Pollen Trap

Harvesting is done by using a pollen trap. A device incorporating a hive's entrance in which incoming bees must pass through two parallel grids of suitable mesh and the pollen loads on the bee's hind legs will knock off and fall into a collecting tray below. Brood rearing would cease inside the hive if the pollen trap stays more than two to three days.

Most pollen traps in sale are fitted at the bottom of a hive (and must have the same cross-section), either immediately above the normal floor board or instead of it.

Harvesting is done by using a pollen trap. A device incorporating a hive's entrance in which incoming bees must pass through two parallel grids of suitable mesh and the pollen loads on the bee's hind legs will knock off and fall into a collecting tray below. Brood rearing would cease inside the hive if the pollen trap stays more than two to three days

Honey and beeswax are the most commonly harvested hive products. Pollen compared with honey has a high protein, vitamin and mineral content. In some advance countries it is harvested and processed for sale.

This is another form of trap to collect pollen from bees; it is simple and easy to collect the trapped pollen.

Tool Box

Beekeeping equipment needs/requirements vary with the size of your operation, types of beehive use, numbers of colonies manage, and the type of beehive products you plan to produce.

Every beekeeper needs a tool box to contains all the necessary tools for the work, ranging from a hammer, knife, plier, nails, bee brush...

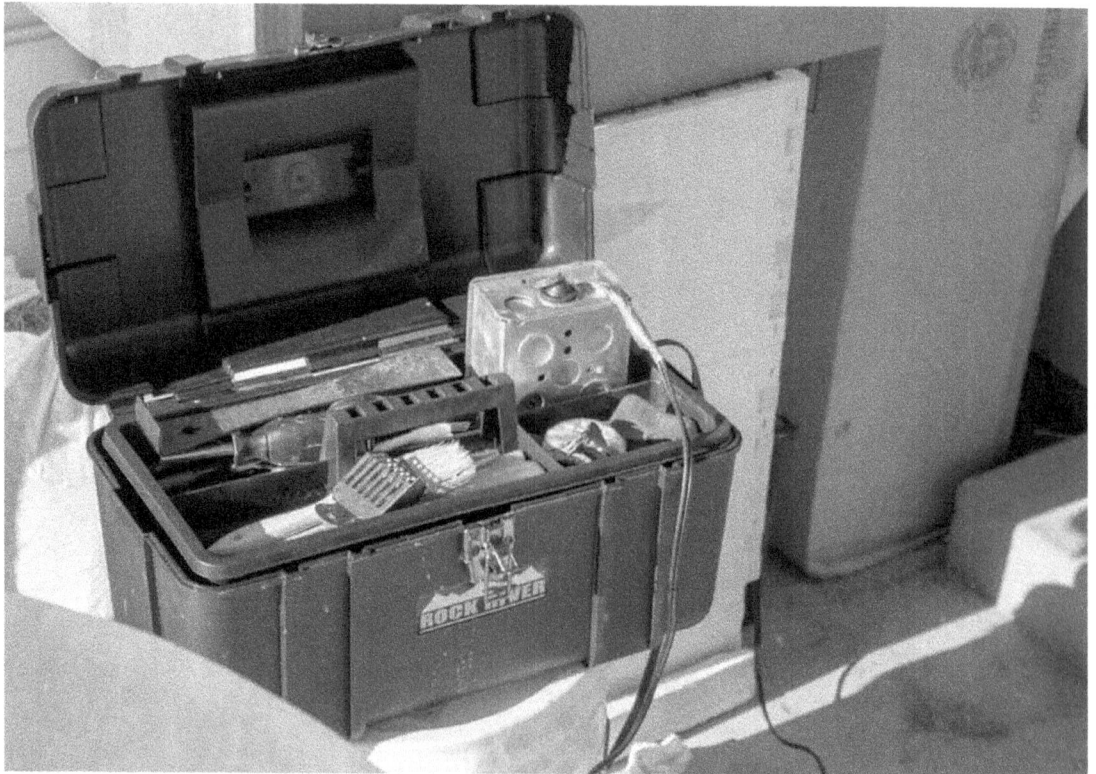

...spur embedder, eyelets punch, straps, queen marking numbers, grafting tool, queen cage, gloves, hive tool, lighter, screw driver, frame grip and much more.

Frame Wiring

Part of the tools use in wiring of frames included: a wiring frame board, frame wire, eyelets, frame wire crimper, eyelet punch, nail, knife and hammer.

Fixing of Foundation Sheet into the Frame

The types of beehive products to harvest determines the kind of equipment to use.
Using a raw natural beeswax sheet as a foundation sheet means you are interested in producing more of beeswax as one of the products.

Unlike plastic foundation sheets. Steam wax extractor machine can be used to melt your old beeswax foundation frame to get more wax.

Tools - Strap

One of the necessary tools for beekeeping is the Strap to hold your beehive together whenever there is need to transport your beehive and any other items in your truck.

Using a strap is more convenient with a Langstroth hive over the Top-bar hive, because the hive features are made in such a way that they interlock without any opening for the bees to come out. You will only need to close the entrance with a small cloth.

Chapter 3 – Handling Bees

Wild Honey & Bee Hunting

Human's first exploitation of bees was through honey-hunting for the sweet, delicious treat of honey. Cave paintings bear testimony to the way honey was collected and similar honey-hunting methods are still used today.

Source: (picture) Smithsonianmag.com.

In the wild, bees will establish themselves in any suitably sized cavity which could be in a hollow tree, a rock crevice, house roof-top or a hole in the ground.

Such a home will be at a favored height, suitably sheltcred, dry, and well ventilated so as not to get too hot or too cold. The entrance will be small enough to be easily defended.

The honey-hunters, who were so skillful as to have discovered the site during the day time, arranged to go back to the feral colony at night to harvest the honey by using naked fire and an axe to cut down the tree to harvest the honey and kill the bees with naked fire.

What to Do to Harvest in the Wild

In developing countries, especially in Africa, because of the nature of their bees being very aggressive and deadly. Beekeepers must be heavily kitted wearing thick underwear together with a bee-suit and armed with good smoke prepared for the harvesting operation. This preparation reduces the attack of bees. Whether the feral colony of the bees is in the rooftop of a house or in the hollow of a tree, a beekeeper must be ever ready to withstand the attack of the bees.

Care must be taken before the operation, especially if the location is closer to residential areas. Human beings and animals can be attacked by bees during the operation, notice should be earlier circulated to the neighbors before the day of operation. If the operation is to be in farming location areas, always notify the neighboring farmers to leave the premises before the operation.

Traditional Beekeepers

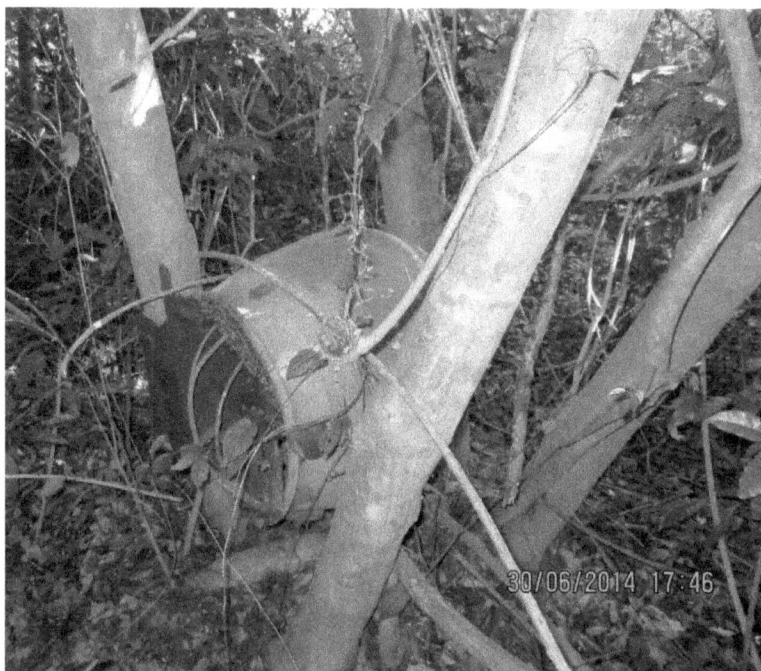

Traditional beekeeping was triggered by accident, as swarms of bees nested in woven baskets or baked-mud water pots left lying around outside in farm settlements. The local settlers quickly took advantage, deliberately providing nest sites for the bees closer to home. Providing a home was far more reliable than hunting for them. Colonies of bees kept in such containers were left alone by the beekeeper.

Traditional beekeepers are often highly skilled knowing exactly where to place hives and when to harvest them; on dark, moonless nights. Both the gourd, straw and log hives are baited and suspended from a branch of a tree. The pot and basket hives are baited and placed on a flat object, like on a large stone or rock; a log of wood to supports it.

The traditional hives are simply made from any suitable material easily available in the area. In many traditional hives, there are no fittings, such as frames associated with modern beekeeping. Bees secure their combs to the interior of the hives; these hives are now referred to as 'fixed comb' to differentiate them from movable comb hives.

The Modern Beekeeper

Modern beekeeping is the art and science of rearing bees, knowing their biology, social behavior, the weather, as well as the use of suitable equipment to maximize the production of honey and other beehive products.

Installation of Package Bees

Tools needed: Beehive, hive tool, sprayer (sugar water mixture), plier, and Bee-suit.

ACTIVITIES:

Suit up! Always wear your veil and glove.

Spray the bees with the sugar solution, this causes them to cluster together.

Use hive tool to remove lid from package. Remove the lid and use the towel to cover the bees so that they will not fly upwards toward the face.

Remove the QUEEN in a queen cage, use a towel to close the hole from where you took the queen cage. Lay the queen in a safe place near your hive.

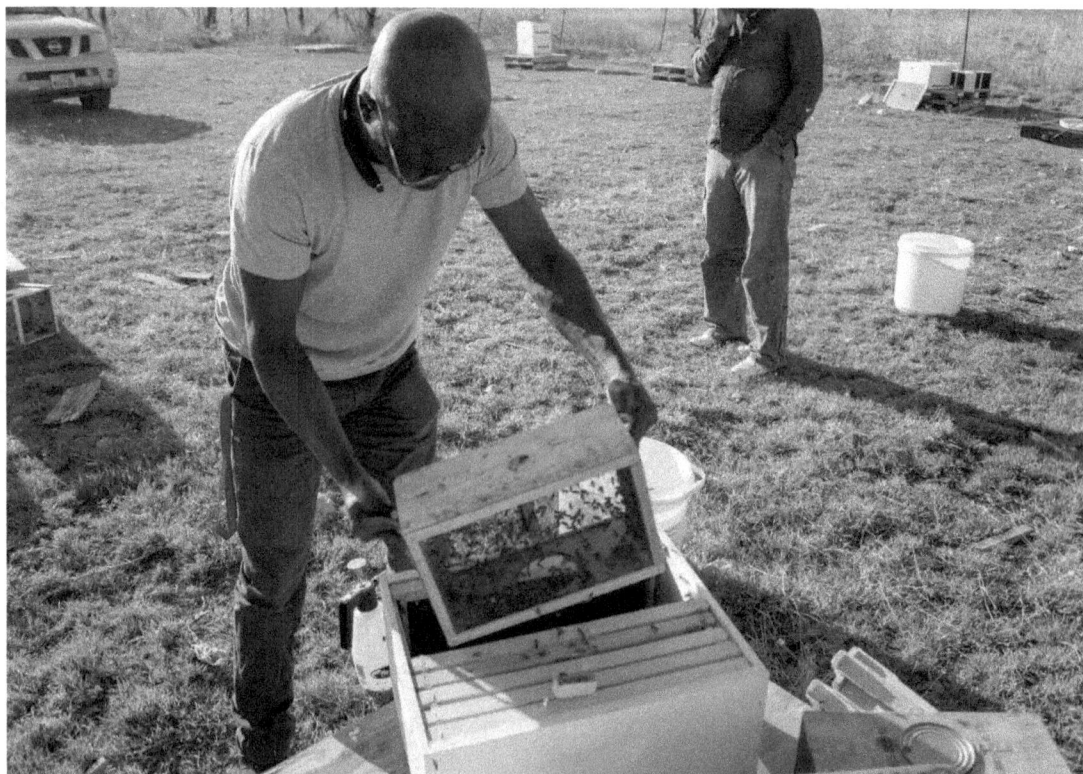

Set the bee package beside the hive, remove outer cover, inner cover; 4-5 frames from your hive, lay to the side. Check the queen to make sure she is still alive.

With your plier, remove cork from the candy end. Make sure it is the candy end to ensure a slow release.

Now, we are going to use copper wire or anything to hang the queen cage as a hanger so that the queen cage will not fall right inside the hive bottom. Spread out a few frames and insert queen candy side up, pushing together frames to give her a snug fit.

Time to install the bees. Tap the bee box to move all bees to the bottom and spray the bees with sugar water. Turn your package bee box upside down over the hive where you removed the 4 frames. Shake to one side, then to the other. Keep shaking. You may give a little tap. Most of the bees should exit the cage.

Place the box at the entrance of the hive and insert your frames back into the hive. Put your inner and outer cover back on.

You have installed your package bees.

Feed the bees immediately.

Package Bee Division

3-pounds package bees can be divided in to two (2) if you have ordered for additional queen together with the package bee (2 queens in one package bee in different cages).

Throw the whole 3 lb. package bee inside a big bowl. Spray them with water to get them wet. Use a big cooking spoon to collect some of the bees in to the opened hive. Insert the queen cage and cover the hive. Feed the bees with sugar syrup immediately.
 Take the remaining bees in the bowl and throw them in the second hive a little distance away facing different direction.

Feeding of Sugar

Feeding bees is very crucial when starting with new Package bees, when winter stores have been consumed, and if honey flow is poor in the fall. Regular monitoring of the hive's condition is therefore very important to the survival of your hives.

A feeder will work well if:

The syrup is readily accessible to the bees, especially in the cold weather; the further bees must travel for syrup the less it will be utilized.

The feeder does not leak.

There is no opportunity for bees to get into the syrup container. Bees will drown if they must access to an open container of syrup (or honey).

There is no opportunity for robbing to start.

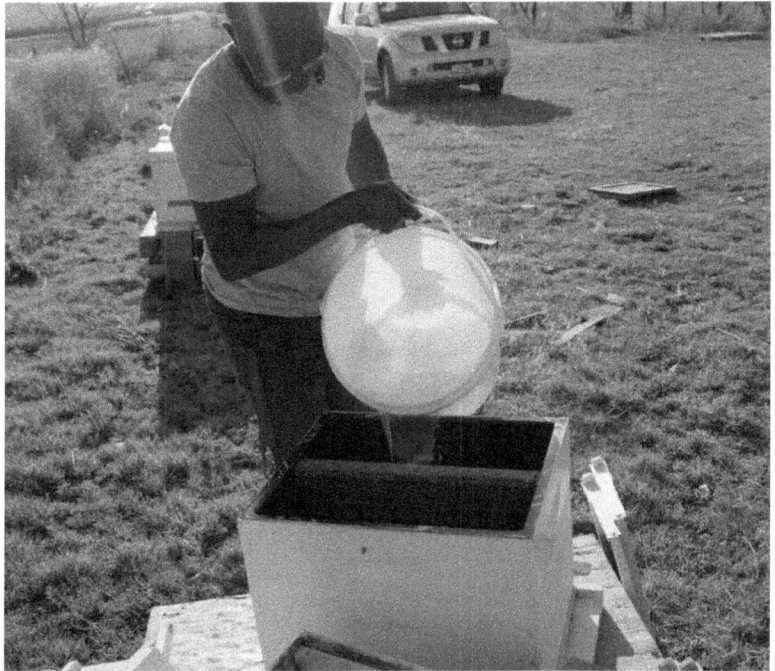

Why Do We Feed Bees with Sugar?

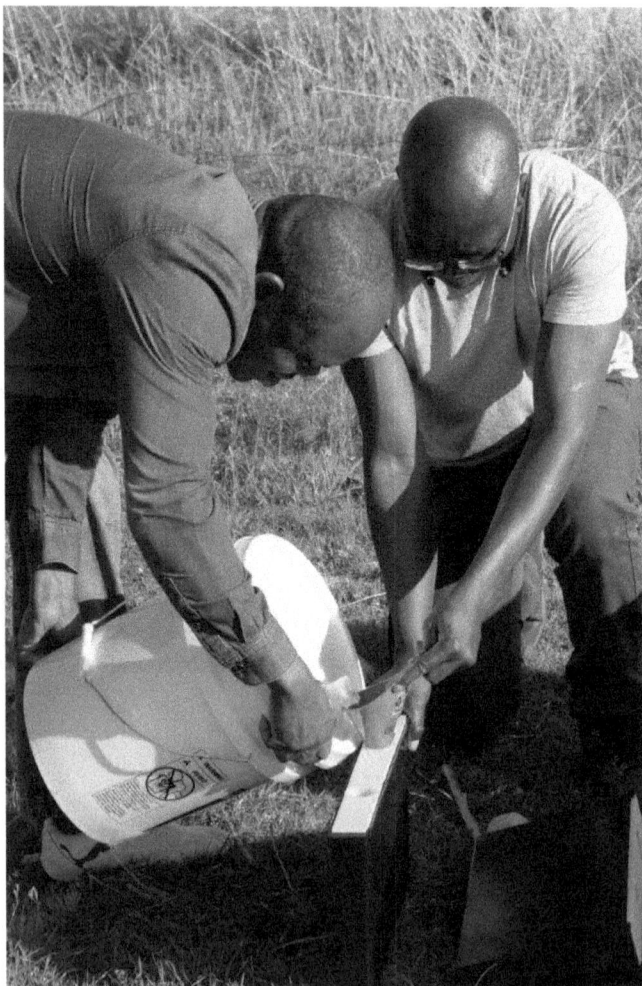

To prevent starvation when colony is about to succumb(rapid).

To enhance wax production and drawing of foundation comb (slow or rapid) depending on the circumstances, e.g. a swarm on foundation is fed rapidly).

To stimulate the queen to lay (usually slow feeding).

As a means of administration of drugs (generally rapid feeding).

To provide adequate store for winter (rapid feeding).

Source: Beekeeping Study Notes ---J.D. & B.D. Yates.

The Precaution to Take When Feeding Honeybee Colonies

Precaution should be taken to prevent robbing (reduced entrances and bee tight hives).

There should be no spilling or dripping of syrup anywhere in the apiary.

Feeding should only be administered in the evening just before dark.

No sugar syrup should find its way to the supers and be mixed eventually with honey for extraction and sale.

Only pure white refined granulated sugar should be used.

Source: Beekeeping Study Notes ---J.D. & B.D. Yates.

Preparing Syrup for Feeding

There are two (2) types of mix, a thick syrup for autumn feeding which will be stored immediately and thin syrup for spring or stimulated feeding which is to be consumed without storing.

Thick ------- 2 lb. sugar to 1 pint of water gives 61.5% sugar concentration.

Thin --------- 1 lb. sugar to 2 pint of water gives 28.0% sugar concentration

Medium --- 1 kg. sugar to 1 liter of water gives 50.0% sugar concentration

Source: Beekeeping Study Notes ---J.D. & B.D. Yates.

Types of Feeders

Pollen syrup feeder: This feeder includes a 4-3/4" super, heavy-duty one-piece plastic liner, and a safety screen to help prevent drowning of your bees. It has a 4-gallon capacity.

2 Gallon Bucket feeder: This 2-gallon bucket feeder is a favorite among beekeeper hobbyists. This easy-to-use design helps protect syrup from robbers to keep your bees fed when outside nectar sources are scarce or when installing new package bees.

The outer cover of the hive is cut opened for the syrup from the bucket to be dropped through a 5-6 punch opening in the bucket for the bees to access the sugar syrup.

Types of Feeders – Dummy Feeders

Division Board feeder: Can be either a deep or medium size. This feeder fits inside your brood box and takes the place of one frame. It holds about 3 or 2 quarts of syrup.

Hive Inspection

Good beekeepers go to a hive for an inspection with a purpose: to take note of the populations growth, to monitor the queen's progress, to assess for disease or illness, and to keep an eye on honey production. There are several clues to look for that tell us what to know.

Source: (urban-livestock/bee-keeping/harvest-honey.aspx).

Do You See a Queen?

The first sign of a colony's health is the QUEEN. Simply put: Is she there and alive? Spotting a queen can take a bit of skill and practice. The healthy, mated queen spearheads the colony, and without her, the colony will not survive unless provided with a new queen. If you search the hive and find the queen healthy and active, your colony has already passed the first test. If you search the hive top to bottom and still can't find her, the next question to ask yourself is…

Source: (urban-livestock/bee-keeping/harvest-honey.aspx).

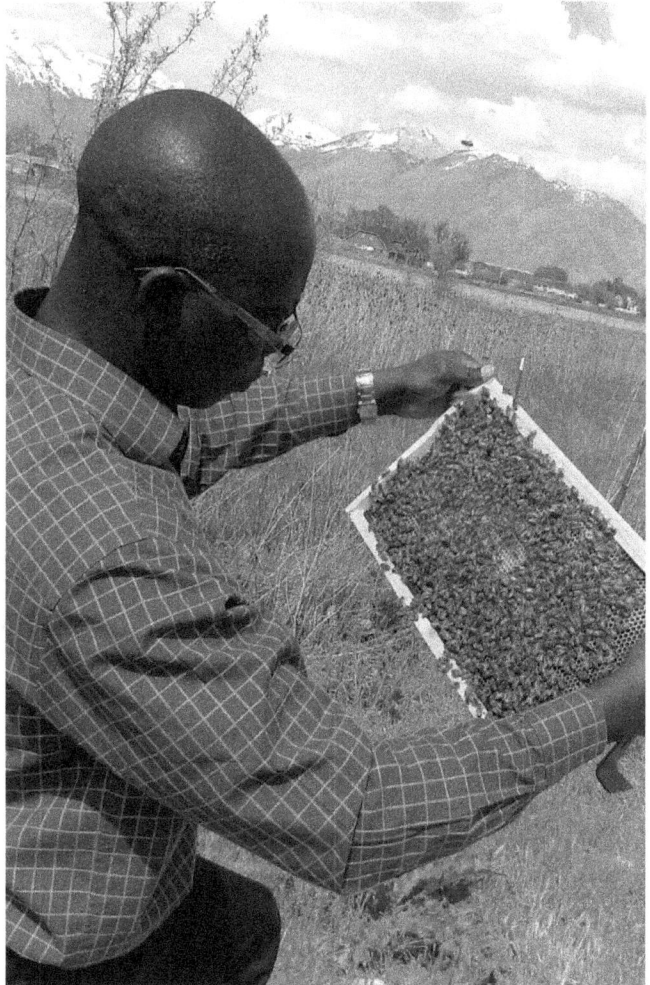

123

Do You See Eggs?

honey.aspx).

Honeybee eggs look like tiny grains of rice floating upright at the bottom of a cell. If you see eggs laid in a healthy pattern — only one per cell, located in the brood nest — you can rest assured that a queen was present in the hive within the last few days.

Sometimes, eggs can be as hard to find as the queen herself. Time your inspections during the mid-afternoon, when the sun is at an angle, and hold up a frame of brood with the sun shining over your shoulder. You might have some luck seeing those tiny grains with just the right light.

If you can't find or see any eggs, ask yourself:

Source: (urban-livestock/beekeeping/harvest-

Do You See Larvae and Brood?

Without finding a queen or seeing eggs, your colony might be on the verge of distress, but don't panic just yet. Look for larvae and brood, clues that colony is growing, and determine their stage of development. Study the stages of honeybee larval development, so you can more accurately determine how long the hive has been without a queen (if it has) and what you need to do about it.

Source: (urban-livestock/bee-keeping/harvest-honey.aspx).

Do you see Supersedure or Queen Cells?

Peanut-shaped queen cells are constructed by worker bees when they know the colony is without a queen or when they intend to replace an old or failing queen through a process called supersedure.

Healthy hives often know before you do if they need to raise a new queen or replace an old one. If you see these cells, you can choose your next move: You may let the bees raise their new queen – or you can cut out the cells and provide the colony with a purchased, mated queen of your choice.

Source: (urban-livestock/bee-keeping/harvest-honey.aspx).

Do you see Swarm Cells?

Swarm cells look just like supersedure or queen cells: They are large (about 1 to 1.5" long), peanut-shaped and constructed to rear queens. However, they differ in intention: Swarm cells are created when the colony intends to cast off a swarm. Typical queen cells and supersedure cells will often be in the center of a frame, while swarm cells will be located along the bottom of a frame.

Source: (urban-livestock/bee-keeping/harvest-honey.aspx).

What does the Brood Pattern look like?

If you have a healthy queen, her nest should display a fair amount of brood in various larval development. Some cells will bulge slightly, which are constructed to house developing drones: male bees that are slightly larger than female worker bees.

A healthy frame of brood looks like a bull's eye, with capped brood in the center and a ring of pollen and capped honey around the outside.

Source: (urban-livestock/bee-keeping/harvest-honey.aspx).

Are the Bees Storing Enough Honey?

The amount of honey a colony needs going into winter will vary from region to region, from hive to hive, and even from season to the next. Keeping notes on each of your hives' inspections throughout the season is a wise move.

Track each colony's health and behavior patterns, local blooms, weather anomalies, and day temperature during each inspection. This practice will improve your beekeeping skills and monitor your bees' progress, making all your future beekeeping decisions conscientious and well-informed.

Source: (urban-livestock/bee-keeping/harvest-honey.aspx).

Colony Manipulation

To manipulate bees successfully, an experienced beekeeper should know what's happening in the beehive in every season of the year. With your experience, you should know what bees are doing at any given time and what you as a beekeeper should do at that time.

The activities you observe in the beehive tell you what to do, or rather, whatever activities you perform in the beehive tells the bees what they should do.

2010.01.05 10:45

Success in beekeeping depends upon a proper exercise of the knowledge of colony organization, growth and behavior in relation to environment as affected by seasonal changes, and the occurrence of nectar and pollen bearing flora.

The beginner should understand that bees can work better when weather conditions are right. The day should be warm, the sun shining, and time selected for the manipulation between 10 o'clock in the morning and 3 o'clock in the afternoon, avoiding quick, and jerky movements.

With the judicious use of smoke, experienced beekeepers can handle bees at any time under practically every condition. But even then, veterans should endeavor to work to their best advantage. The beginner should select his time, and endeavor to make his movements very deliberate, again, avoiding quick jerk movements.

What to do for Colony Manipulation

It is expected of the beginner beekeeper to work from the side of the hive and never from the front so that the flight of bees will not be obstructed either with the beekeeper's body or with the hive cover, when frames or supers are removed during examination. Have the smoker lit and operating so that there is always available a large volume of dense smoke.

A heavy smoke may occasionally be needed but in most cases, a very light smoke or even no smoke may be what is required. Remove the outer cover, insert the blade of the hive tool under one corner of the inner cover, raising the cover slightly, just enough to smoke into the opening. Wait at least 30 seconds before removing the inner cover.

A first frame must be removed to provide space for removing the others. As each frame is removed stand them on end on the ground or on the end of the covers, leaning them against the hive. Place the frames near the front of the hive away from underfoot. If for some reason the bees or the queen does not crawl or fall from the frame, they will be near the entrance.

Examination of the brood nest are necessary to maintain a continuous check on the presence of a queen, a condition that is easily determined by checking for the evidence of eggs, larvae and sealed brood. Replace the frames in the same order that they were removed.

2010.01.05 10:45

Working with a Top-bar hive is so exciting and not as hectic as using a Langstroth hive because there is nothing to lift except the frames to inspect.

In Top-bar hives the frames are arranged side to side, and relatively closer to each other. For this reason, it is advisable to start working from the back of the hive to remove the last frame. The bees will commence making of the combs from the front of the hive close to the entrance hole.

When you remove the last frame bar in the hive, there will be space to drag the remaining frames backward to inspect them.

Moving of Hives

Many factors can prove responsible for the movement of beehives from one location to another. These include:

Lack of adequate supply of natural resources, examples nectar, pollen in the area.

The land owner may no longer want you at the location because of your bees' continual nuisance, or perhaps a fight between the two parties necessitates movement of the beehives.

Natural dangers threatening your bees may be the reason for movement such as bears or insects.

Moving/Loading Hives

Before loading a colony, you must close the entrance with the reduced entrance tool and cover the small opening with cloth. The opening will be opened immediately getting to your new location while you station the trailer with the colonies right inside it.

Moving colonies on the farm can be so tedious and worrisome, use of this equipment makes it so much easier to carry 3-4 hives at a time to load them into a truck.

Working with bees can be hard labor. Honey is heavy couple with the weight of the beehive. Having a suitable tool for the beekeeping activities can make the work more interesting and rewarding.

A trailer loaded with colonies coupled to a bus for moving colonies from one location to another can really ease the stress.

Tools to Transport Beehives

Bees can be transported in this open collapsible trailer.

It can serve as a mobile bee house that can be stationed at a place.

Moving Bee Colonies

The colony loading operation must be done at night a day before the movement. The colony should be properly arranged when all the bees have returned to the hive. Close the entrance with the reduce entrance tool and cover the small opening with a cloth. Open the entrance immediately arrive at the new location and station the colonies.

Moving Bee Houses

Bee house offer security against vandalism and theft of hives, and the hives are protected from the harsh weather and therefore last longer.

The Bee house is portable and can be taken down, moved about, and relocate to another location.

The bee house contains almost 12-15 colonies in both sides of the house, as well as super boxes and other tools can be kept inside it.

Movable Bee House

Mobile bee houses can be permanently stationed in one place for pollination services.

The advantage of the mobile bee house is that it can be coupled to a truck and can be carried to anywhere anytime.

Station Bee House

The station bee house contains enough space to occupy a few beehive colonies, other necessary beehive tools, a small lecture classroom with sitting facilities, and a location for fixing beehive components.

The only disadvantage is that it is not movable, and the problem is that if there is not enough forage around for bees, nothing can be done to solve the problem.

The bee colonies entrance faces the east side of the building for the rays of sunlight to enter.

Mini Station Bee House

This is a mini station bee house for a few beehive close to the house.
 It can contain 3 -4 colonies. It keeps the beehives away from direct access by the neighbors and preserves the beehive from direct harsh weather.

Chapter 4 - Harvesting & Extraction of Beehive Products

Harvesting & Extraction of HONEY – Bee Escape

Preparation for the honey harvest.

When the honey is due for harvesting in the hive, one tool to use to keep bees away from the honey frame is called BEE ESCAPE. It should be installed in-between the super and brood box the evening before the operation. When the bees in the super go down the hive, they will not have access to get back to the super box, and the few bees remaining in the super will be removed by brushing them off with bee brush or by use of a bee blower machine.

Harvesting & Extraction of HONEY – Bee Blower Machine

Using a bee blower machine to remove the bees in the frames into a container to get the bees free of a super frame.

For a small-scale beekeeper, a bee brush is a good tool to get the bees away from the frame.

Harvesting & Extraction of HONEY – Bee Escape

The process of harvesting honey frames starts with the use of bee escape to prevent the bees from getting back to super hive. The bees are collected into a yellow plastic swarm box to establish more mating nuc hive.

Honey Extraction – Uncapping Machine

The honey frame can be uncapped with a honey fork, or with this mechanical uncapping machine. Get the honey frame ready for the centrifugal honey extractor. Spin the honey out of the comb.

Honey Extraction – Honey Fork

The harvested honey frames are taken to a honey house where the honey will be extracted. The honey fork is using to open the capping wax that the bees used to cover the honey frame before putting the frames inside the centrifugal honey extractor.

Honey Extraction – Honey Press

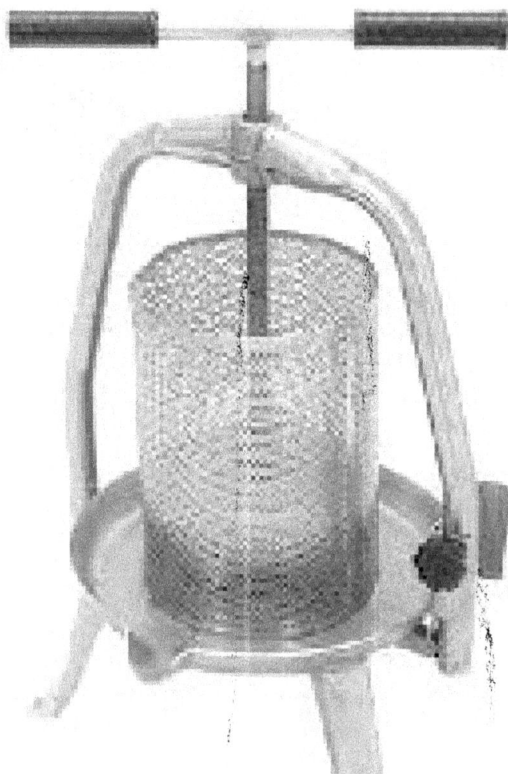

A honey press is an ideal tool to extract the honey from a Top-bar hive. Honey comb is cut from the bar when the honey is ready for harvesting and put in a straining bag. Right away it is placed in the honey press bucket. The rod is screwed downwardly to press the honey out of the comb inside the straining bag.

The honey is then collected through the receptacle bowl below the straining body pouring directly to the honey bucket below.

Source: (picture) paynesbeefarm.

Honey Extraction – Centrifugal Honey Extractor

This is the tool used to extract honey from the frames. It operates by centrifuging the honey out of the combs and into the wall of the extractor at a high speed, where the honey falls by gravity to the bottom.

The temperature of the room is important when extracting honey from the combs because honey flows very well quickly when warm than when cold and will flow less if left in the cells of the comb too long.

Extracted Honey – Honey Syrup

Honey is the natural substance produced by the honeybee.

Extracted Honey is honey only obtained by centrifuging de-capped brood-less combs with or without the application of moderate heat.

Pressed honey is honey obtained by pressing brood-less combs with or without the application of moderate heat.

Drained honey is honey obtained by draining de-capped brood-less combs with or without the application of moderate heat.

Source: Beekeeping in the United States.

After Extraction of Honey from the Frames

What do you do after the honey has been extracted from the frames?

Return the super hive with the honey-wet back to the bees. Store the dry frames for use next year.

Honey House

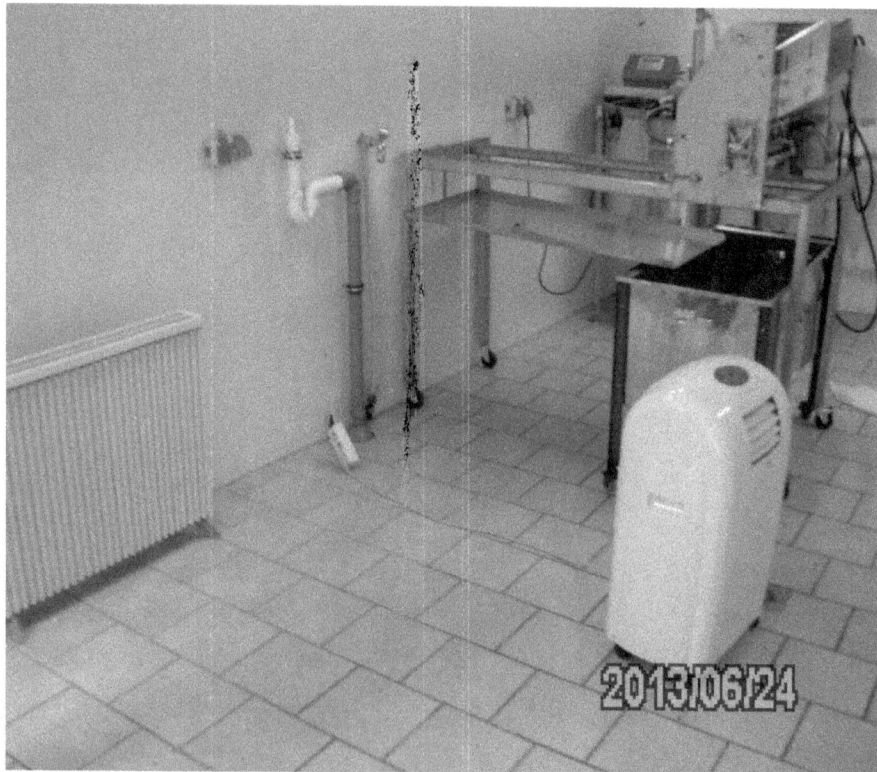

A honey house is any stationary or portable building or any room or place within a building, including its equipment which is used for the purpose of extracting, processing, packing, and/or other handling honey.

There is a guideline for honey house sanitation code as adopted by the American Beekeeping Federation ABF which must be strictly observed.

Source: www.abfnet.org.

Honey House – Honey Storage Tanks

Among the items to be in the honey house are as follows:

Honey extractor	Uncapping machine
Uncapping honey fork	Refractometer
Uncapping knife	Stainless steel uncapping tank
Honey strainer	Capping spinner
Honey storage tanks	Stainless steel double sieve

Bottling & Packing of Honey

Those areas that are used for the filling of honey and storage of clean containers must be protected from contamination of all kinds.

Containers must be approved for food use and must be adequately cleaned prior to filling to remove all traces of rust.

After filling, containers should be protected from direct sunlight and excessive heat to avoid deterioration of the honey.

Containers should not be stored where dust, rain, or other foreign material might enter the container through a loose or partially sealed lid.

Beeswax Extraction & Processing – Beeswax Capping

Beeswax capping is removed when honey is to be extracted. You will need to get the honey frames ready for honey extraction, and then, put the capping into the capping spinner to remove the pure clean beeswax.

Beeswax Capping Spinner Machine

This beeswax capping spinner machine is great for those with 100 beehives or less. This machine separates your honey and beeswax – capping is easy.

Put the wax capping on the spinner machine. Cover the machine for a specific length of time. All the wax will melt and separate your liquid wax from honey.

Steam Beeswax Extractor

As the name implies, the machine uses steam to melt the beeswax combs arranged inside the box. The machine has a component that contains water. When the temperature of the water gets to a boiling level, the steam melts the combs and wax is collected through the opening tap with water and wax collecting when solidified.

Steam Wax Extractor

This is a steam wax extractor in a rectangular shape. The machine contains as much as 15-18 old stored comb frames. The wax is usually collected with water in a basin/bucket place underneath. The wax collected always comes with dirt and impurity, which will later be cleaned to get pure, along with the clear beeswax.

Cleaning of the Frames

Sterilize your frames using washing soda crystals. Boil the water to the boiling point. Add some quantity of soda crystal to it and submerge your frames in the boiling soda for a few minutes until the frame is clean of any remaining wax or propolis.

Remove the frames. Rinse them down with water to remove residual sodium carbonate and then leave them to dry, ready for re-use.

Cleaning of the Beeswax

When the beeswax is first extracted from the frames it contains dirt, impurities, and some cocoons of the brood.

Cleaning of the dirt off the wax is of paramount importance. Immerse the wax in boiling water. Pour in a bag which has been put under a press to draw out the pure clear beeswax. The remnants of the dirt remain in the bag. While the wax in the water left to cool, it solidifies and collect, and become pure wax.

Pure Beeswax Extraction

Using a beeswax capping extraction machine separates your wax and honey in good condition.

Beeswax for Foundation Sheets

When it is time to use beeswax cake for foundation sheet. Soaked it in hot water to make it soft for the machine. A clean beeswax cake be formed in any shape depending on the mold used. Melt the wax in water-immersion to get liquid wax. Pour the liquid wax into a rectangular mold as shown in the picture.

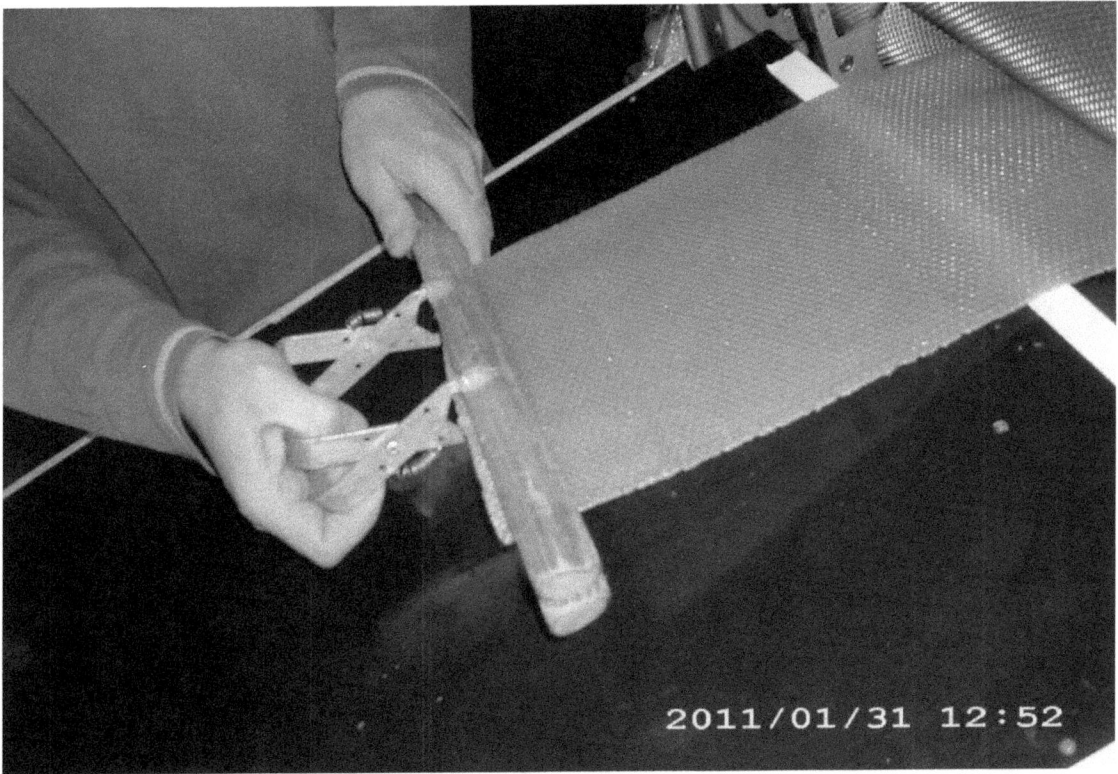

The beeswax is now in the foundation sheet mold machine. To emboss the wax with the honeycomb sized to assist the bees. The bees will start drawing out comb cells from the already embossed foundation sheets.

Making of Foundation Sheets

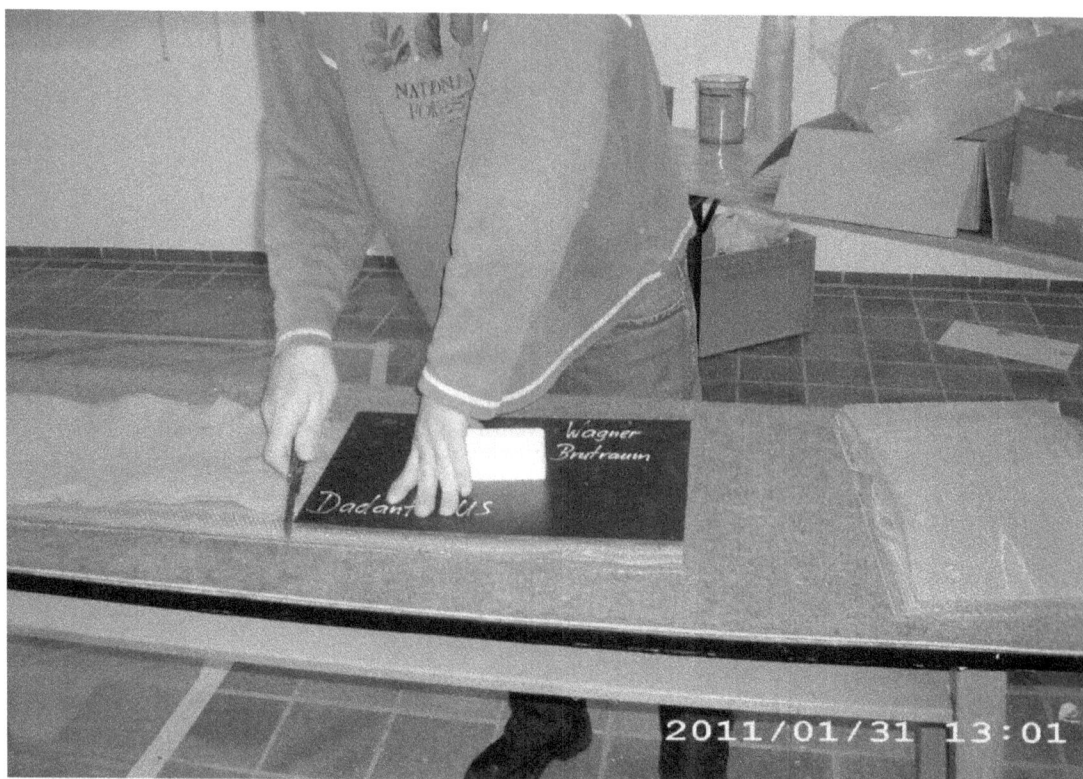

One of the products that can be made from beeswax is the foundation sheets for the frames. It assists the bees to make the combs faster by just drawing out the comb cells.

The foundation sheets can be cut in to sizes of your choice depending on the kind of frame you use, either brood box or super box.

Bee Pollen

Palm tree is one of the bee plants that produce plant pollen, a protein source for the bees.

Bee Pollen Collection

Bee pollen is one of the beehive products derived by keeping bees. It has many nutritional and medicinal benefits to the body. The picture shows the tray for pollen collection.

Bee Pollen

Bee pollen is the food of the young bee and is considered one of nature's most completely nourishing foods.

It contains nearly all nutrients required by humans.

The pollen stored in the freezer can be immediately consume.

Bee Propolis

Propolis is one of the by-products of the beehive. The bees use it to glue the frames together and to close every opening found in the beehive.

You can collect the propolis by scraping the tip of the hive body or by scraping the side of the frames with the hive tool.

Bee Propolis Trap

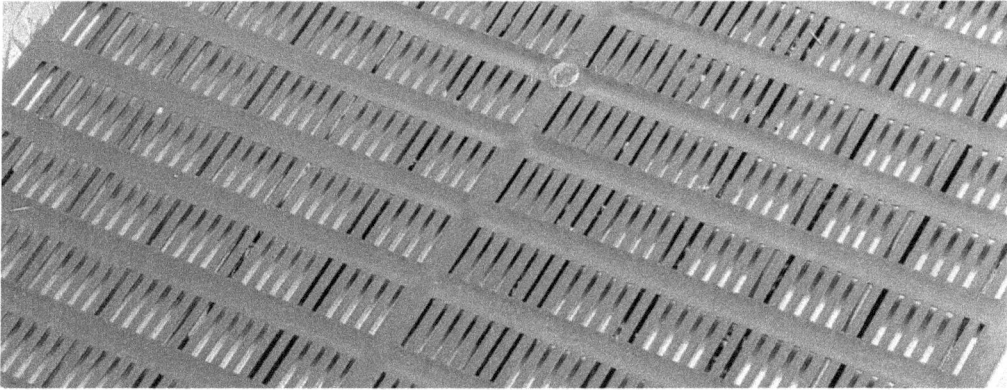

The propolis trap is a simple flat sheet with width slits of 6.0mm that bees will close with propolis.

The tool will be inserted at the top of the hive. The bees will regard it as opening and work to close it with propolis. Remove the sheet and place in a deep freezer. The propolis can released from the sheet by shattering.

Bee Bread

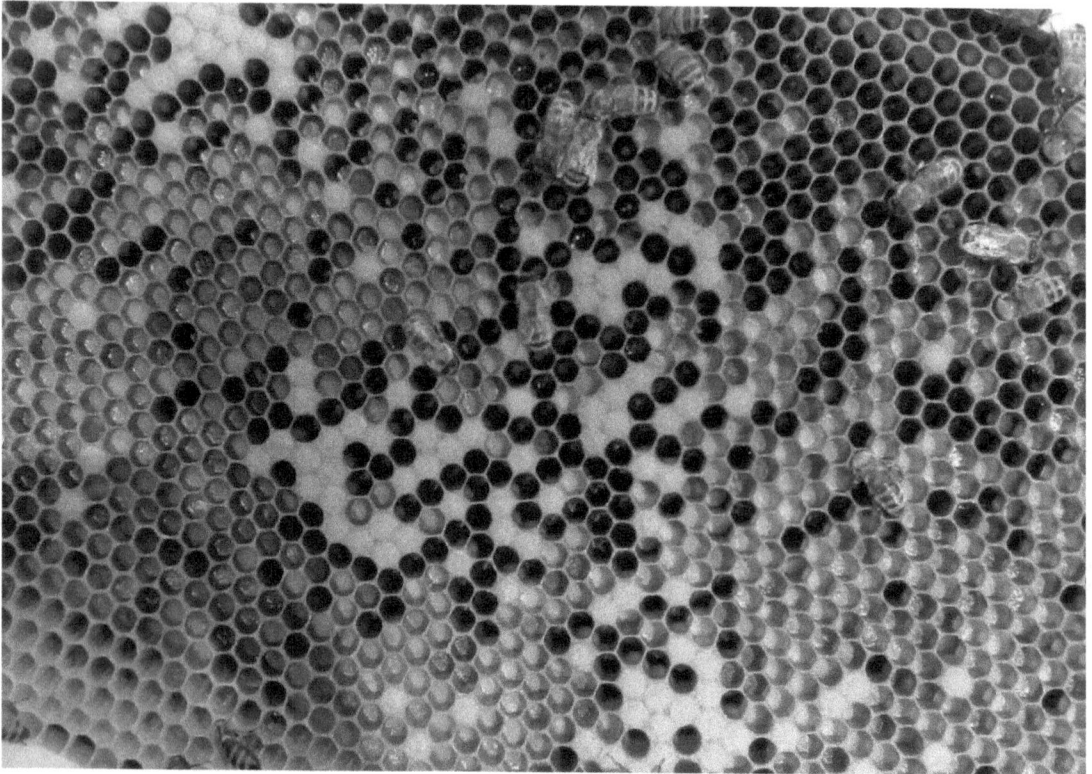

Bee Bread is just one of many beehive products derivable from keeping bees. It is the stored bee pollen inside the comb cells.

Harvesting of bee bread is simple. All you need to do is to cut the comb in which it is stored from the frame.

Harvesting of bee bread is simple. All you need
to do is to cut the comb in which it is stored
from the frame.

Royal-Jelly

Royal-Jelly production can be done during the process of queen breeding, when the bees are sensitized to raising queen cells.

Bee Venom Therapy (BVT)

Bee venom therapy is one beehive product, and it works in the same way as acupuncture medicine.

This therapy is not advisable to employ in isolation. It must be in conjunction with other beehive products and other natural products.

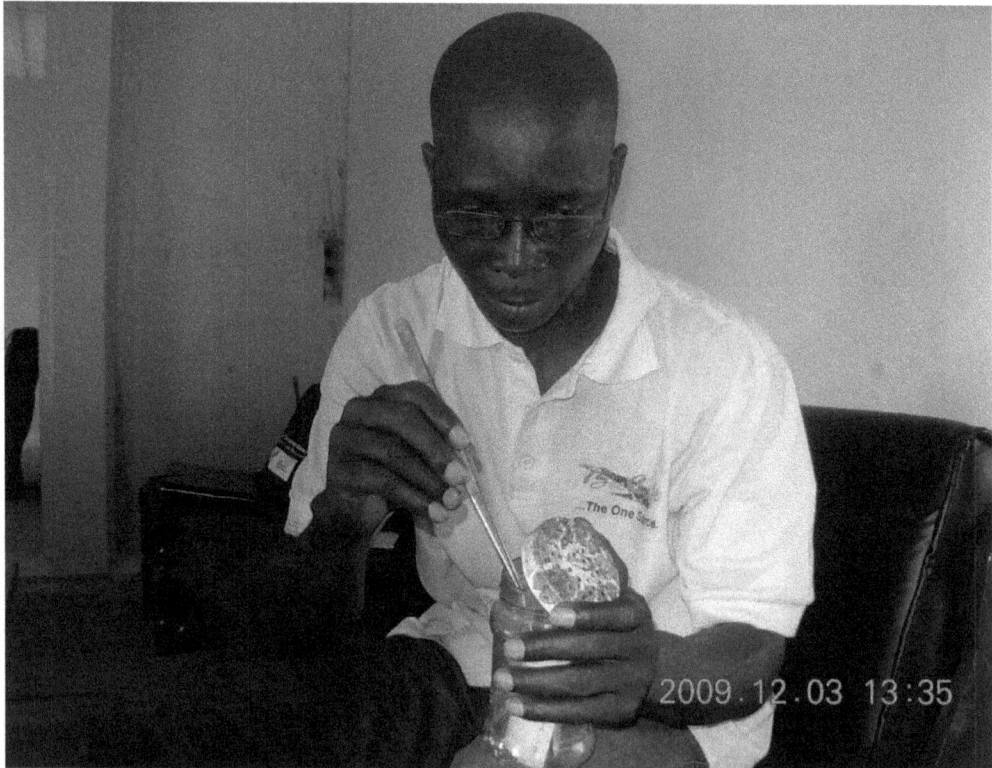

There are rules and conditions to follow before engaging in BVT:

Be a trained practitioner of BVT

Own bee colonies to get live bees.

You must first do a trial test to confirm that the patient is not allergic to BVT.

Start with a few number of bees, between 3-4 bees at the first attempt.

There is a need for training beyond the little information provided in this book.

Apitherapy

Apitherapy is the use of beehive products for the healing of human ailments. These includes honey, Pollen, Royal-Jelly, natural oil (Aromatherapy), medicinal herbs and massage.

Apitherapy is the future of medicine for our WELLNESS.

Chapter 5 – Queen Rearing Techniques

Queen Rearing

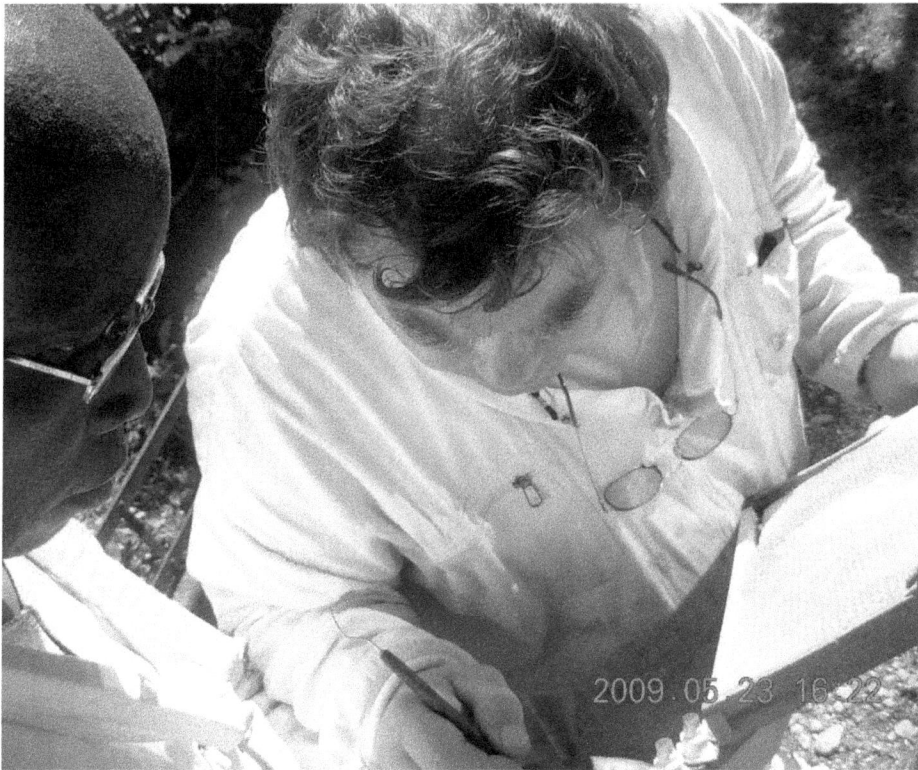

Queen rearing is a highly specialized process and is an essential part of beekeeping.

The tools needed for the rearing of queen bees are as follows:

Grafting tool

Queen cage

Push-in Cell protector

Stainless Steel Queen catcher

Queen Rearing Kits

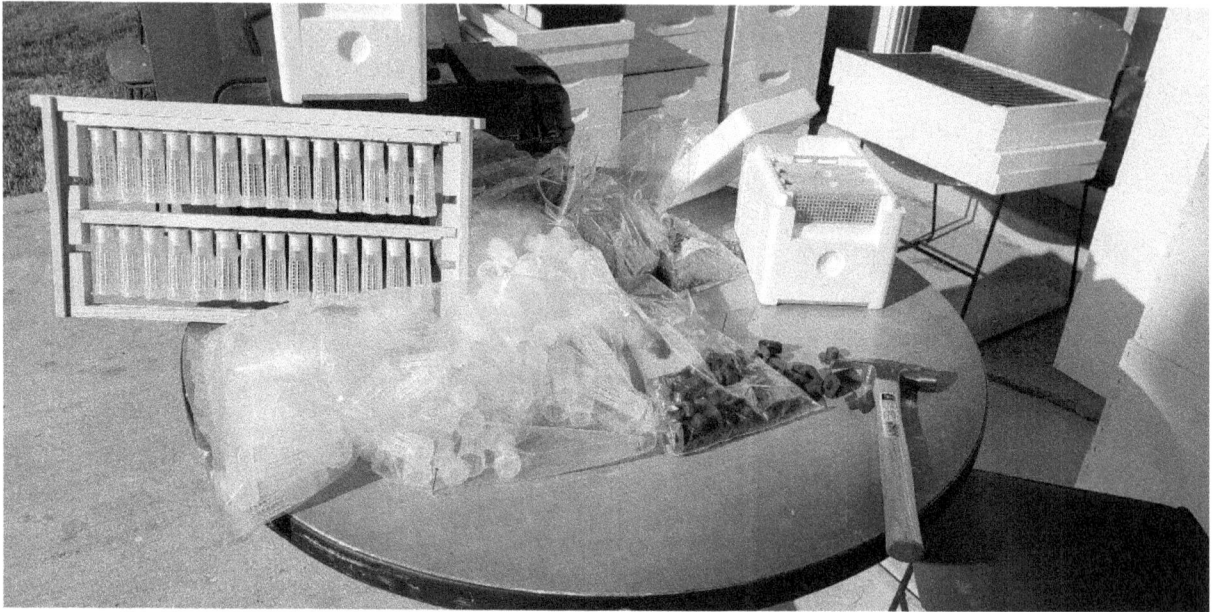

Hair Roller

Brown Cell Fixtures

Yellow Cell Cup Holders

Brown Cell Cups

Mating nuc Box

Corrugated Cardboard nuc Box

Royal-Jelly

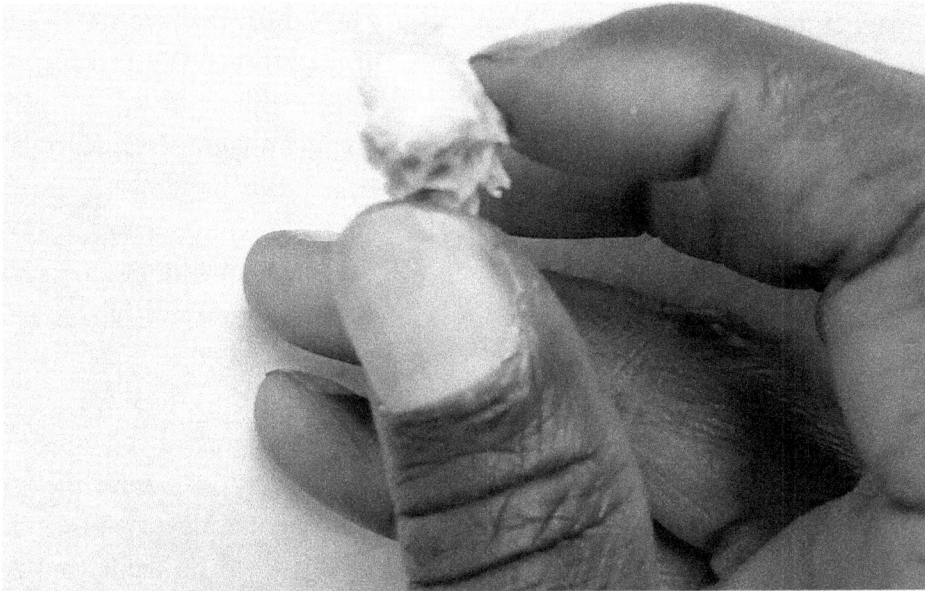

Royal-jelly is food for the queen. You can use royal-jelly if available as one of the items for queen breeding.

Putting royal-jelly first in the cell cup before grafting the larvae makes the bees accept the larvae quickly.

Queen Rearing Techniques

The process requires step by step procedures.

Prepare a queen-less cell builder colony for at least 4 days before introducing the frame.

Grafting of larvae of 24-36 hours old age from the breeder colony with the help of grafting tool in to the cell cups.

Leave the grafted cell cups in the frame in the cell builder colony for the next 4 days before checking the development, that is, if the bees have worked on them, and covered the queen cells.

Remove the sealed queen cells from the frame. Put the sealed queen cell with some attendant bees in the hair roller. Return the frame to the cell builder colony until the virgin queen is released inside the hair roller.

Remove the sealed queen cells from the frame. Put the sealed queen cell with some attendant bees in the hair roller. Return the frame to the cell builder colony until the virgin queen is released inside the hair roller.

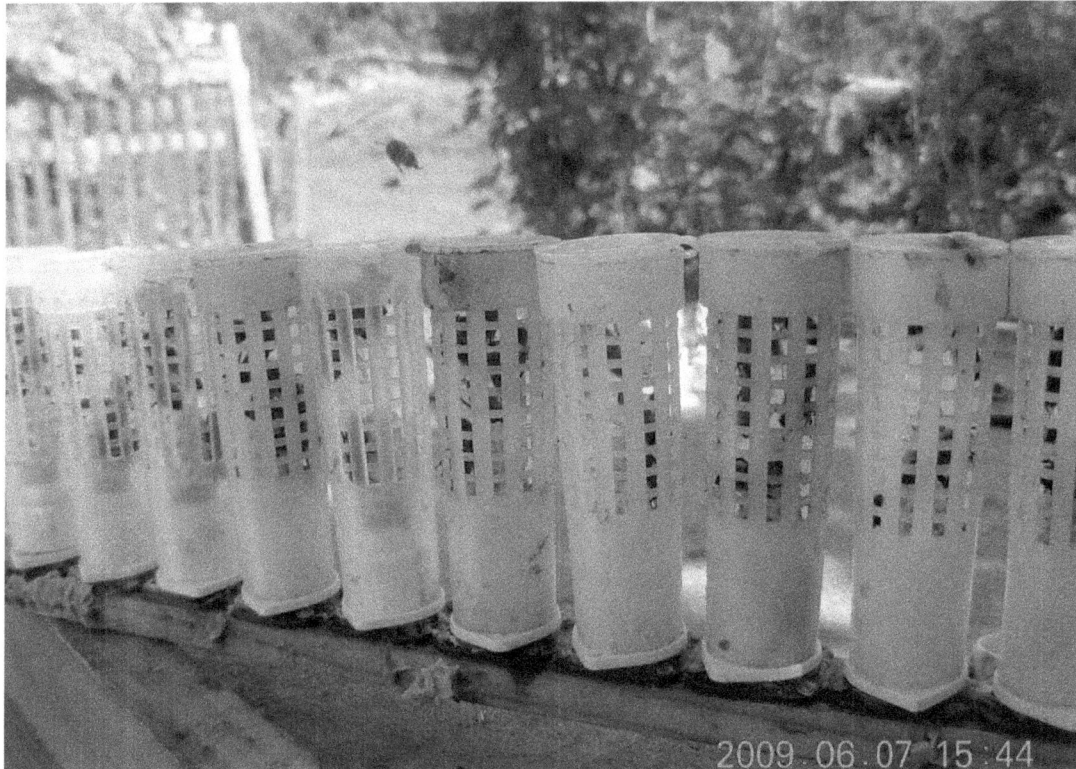

Return the sealed queen cells inside the hair roller back inside the cell builder colony for the next 10-12 days for the emergence of the virgin queen bees.

Next to get is the mating nuc hive ready.

Collection of Bees for Mating Nuc

Collection of bees to establish mating nuc hive for virgin queens to be fertilized during the flight mating:

This exercise can be done after carefully remove the queen from the strong colony to get some bees for mating nuc.

It can also be carried out when preparing for the harvesting of honey when using the bee escape to keep the bees from the super. Those bees left under the bee escape can be collected for this purpose.

Queen Rearing Techniques – Mating Nuc

Next, collect worker bees from another beehive to make the mating nuc for the virgin queen to mate with drones for fertilization. The queen begins laying eggs inside the nuc before taking her to the bigger beehive.

Queen Rearing Techniques

The fertilized queen will begin laying eggs and the colony will start to grow. Later, take away the queen to a normal beehive to continue the colony development.

The mating nuc hive for fertilization of virgin queens.

Introducing Queen Cell into a Queen-less Colony

A queen cell or swarm cell can be introduced into a queen-less colony with the help of a top-bar cell protector.

Take a brood frame along with the young bees to support the emerging virgin queen in the queen-less colony.

Introducing Queen Cell into a Mating Nuc

A queen cell can also be introduced into the mating nuc for the virgin queen to undergo her mating flights for fertilization.

Raising Queens to Increase Your Colony

This method is for a beginner beekeeper who wants to increase the numbers of his/her colony. All it takes is to keep your eyes on the population of bees until the bees produce the swarm cells – when they are ready to leave the colony because of over-population.

Then you divide the colony by removing a frame with the swarm cell, one or two brood frames and some quantities of young bees into the new colony.

Pest and Diseases

Varroa destructor mites treatment:

Varroa mite is an external parasitic mite that attacks the honey bees. It attaches to the body of the bees and weakens the bees by sucking fat from their bodies.

Treatment 1.

Use drone cells to capture the mites; the mites like to reproduce in the drone broods.

Make an empty frame available for the bees. The bees will make a drone comb cell to raise drone bees.

Sources: (Photo) malietahoney.com.au

The mites like to reproduce right inside the drone brood cells, you will want to wait until the brood is sealed/capped and then take them out to give to the birds to eat. You can also put it in the freezer for 36 hours to kill all the varroa mites.

Treatment 2.

Sugar powder treatment: Take 2 tablespoons of powdered sugar in a glass jar. Take a frame with bees. Make sure queen is not included. Shake the frame of bees in to a bowl. Take a few handful quantities of bees.

Put the handful of bees in a glass jar with powder sugar and shake together so that if the mites are on the bees, they will fall in the sugar.

Take the glass jar with both the bees and the sugar, shake it into a plastic bowl so that the powdered sugar might fall into it. If the mites are found on the bees, you will then know the degree of infestation and the treatment that should be applied.

Wax Moth

Wax moth is one of the most serious pests to honeybee colonies. It is not a threat to normal colonies and cannot kill a colony, but weakened colonies invaded can eventually be destroyed.

Combs are most often destroyed by the wax moth when stored in dark, warm and poorly ventilated rooms. However, there can be a considerable damage to combs even while in use especially in hives where the population of adult bees is too small to protect all the combs.

Make sure you combine your weak colony with your strong colony and reduce the flight entrance when the colony has not been strong enough to defend itself.

Use dry tobacco leaves with dry grass as smoking materials.

Chapter 6 – Economics of Beekeeping

Economics of Beekeeping

Lectures on the rudiments of beekeeping among potential beekeepers:

Starting a business from a scratch is a road only traveled only by an adventurous individual. The true entrepreneur seeks a business that is his own creation. This individual requires the satisfaction that derives from the conceiving of an idea and developing it into a profitable business.

Starting a Beekeeping Business of Your Own

Research evidences suggest that the following factors are critical in starting a small business:

*The Market

*Customers

*Need for a Partner

*Start-up Money

*Facilities for the operation (Equipment & Supplies)

*Personnel

*Inventory

*The Business Facilities

*A detailed Business Plan

*Location and

the individual.

Bee Profitable

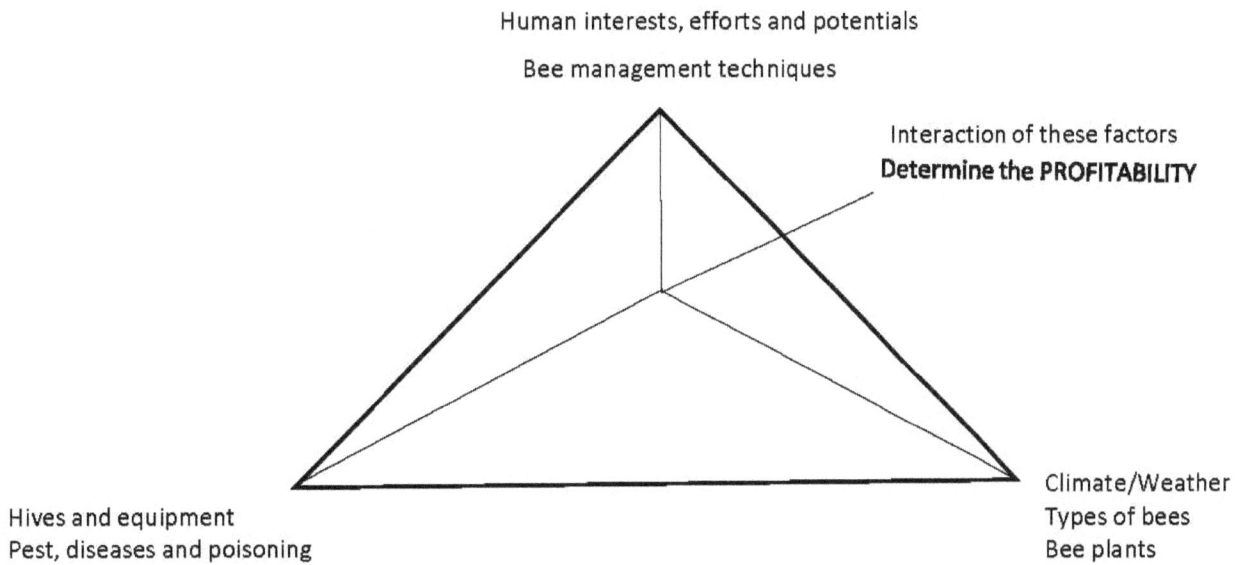

Human interests, efforts and potentials

Bee management techniques

Interaction of these factors
Determine the PROFITABILITY

Climate/Weather
Types of bees
Bee plants

Hives and equipment
Pest, diseases and poisoning

Managing Beekeeping Business for PROFITS

The main limiting factor in creating and keeping a profitable bee business is a human lack of KNOWLEDGE. This lack can be resolved by acquiring education and sharing your experiences among the beekeepers.

This profitable venture can only be achieved when the ecology and biology of bees is understood so that predictions can be made about the behavior, and management techniques can be developed.

The major ingredients to succeed are money, skills and more importantly, internal drive (ability to make the dream a reality).

Project Management App

A Trello board is a list of lists, filled with cards, used by you and your team. It's a lot more than that, though. Trello has everything you need to organize projects of any size. Open a card and you can add comments, upload file attachments, create checklists, add labels and due dates, and more.

Project Financial Management Tool

There's more to your business than accounting; and sometimes, to get the job done, you need extra tools. QuickBooks Apps expand the capabilities of QuickBooks Online, each working together to improve your business.

Adesina Daniel Oduntan

Beekeeping Practice

BEEKEEPING (APITHERAPY) is the future

MEDICINE

WELLNESS &

FITNESS

Notes

Adesina Daniel Oduntan

Notes

Notes

Adesina Daniel Oduntan

Notes

Notes

www.ingramcontent.com/pod-product-compliance
Lightning Source LLC
Chambersburg PA
CBHW080316220326
41519CB00072B/7432